Solutions Manual to Accompany

BIOCHEMISTRY

DONALD VOET
University of Pennsylvania

JUDITH G. VOET
Swarthmore College

WILEY

JOHN WILEY & SONS
New York • Chichester • Brisbane • Toronto • Singapore

Illustration and Production: J/B Woolsey Associates

ISBN 0-471-50242-1

Printed in the United States of America

10 9 8 7 6 5 4 3 2

PREFACE

This book contains the answers to the end-of-chapter problems in *Biochemistry* by Donald Voet and Judith G. Voet. As we stated in the preface of that text, we feel that working out problems is an essential part of the learning process. We therefore urge students to make a serious effort to answer a problem before consulting this manual. Ideally, the answers contained herein should be used only to check the validity of answers in which the student has confidence.

The problems that are answered here have been tested on several generations of Penn and Swarthmore biochemistry students who cheerfully accepted their roles as guinea pigs. The answers were also carefully checked for accuracy and completeness by Dr. Ann E. Shinnar of the University of Pennsylvania and Dr. Nancy Hamlett of Swarthmore College to whom we are grateful. We also thank Delores Magobet for typing the manuscript and John and Bette Woolsey and Patrick Lane for executing the drawings and generating the final copy.

DONALD VOET
JUDITH G. VOET

Chapter 1
LIFE

---

1. The number of cells in 10 L of saturated culture is :

$$10 \text{ L} \times 10^3 \text{ mL/L} \times 10^{10} \text{ cells/mL} = 10^{14} \text{ cells}$$

$$2^n = 10^{14} \text{ where } n \text{ is the number of doublings}$$

$$n = 14/\log 2 = 46.5$$

Since each doubling takes 20 min = 1/3 hour,

Time to reach a saturated culture = 46.5/3 = **15.5 hours**

Volume of an *E. coli* cell = $\pi r^2 h = \pi \times (1 \times 10^{-6} \text{ m}/2)^2 \times 2 \times 10^{-6} \text{ m}$

$$= 1.57 \times 10^{-18} \text{ m}^3$$

Volume of 1 $\text{km}^3 = (10^3 \text{ m})^3 = 10^9 \text{ m}^3$

Number of *E. coli* in 1 $\text{km}^3 = 10^9 \text{ m}^3/1.57 \times 10^{-18} \text{ m}^3 = 6.37 \times 10^{26}$ *E. coli*

$$2^n = 6.37 \times 10^{26}$$

$$n = \log\,(6.37 \times 10^{26})/\log 2 = 89 \text{ doublings}$$

$$\text{Time to reach 1 km}^3 \text{ volume} = 89/3 = \textbf{29.7 hours}$$

2. See Figures 1-2 and 1-5

3. For an *E. coli* cell:

$$\text{Surface area of cylinder} = 2\pi rh + 2\pi r^2$$

$$= 2\pi(1 \times 10^{-6}\,\text{m}/2) \times 2 \times 10^{-6}\,\text{m} + 2\pi(1 \times 10^{-6}\,\text{m}/2\,)^2$$

$$= 7.85 \times 10^{-12}\,\text{m}^2$$

$$\text{Volume} = \pi r^2 h = \pi(1 \times 10^{-6}\,\text{m}/2)^2 \times 2 \times 10^{-6}\,\text{m} = 1.57 \times 10^{-18}\,\text{m}^3$$

$$\text{Surface-to-volume ratio} = 7.85 \times 10^{-12}\,\text{m}^2/1.57 \times 10^{-18}\,\text{m}^3 = 5 \times 10^6\,\text{m}^{-1}$$

For a eukaryotic cell:

$$\text{Surface area} = 4\pi r^2$$

$$\text{Volume} = \frac{4}{3}\pi r^3$$

$$\text{Surface-to-volume ratio} = \frac{4\pi r^2}{\frac{4}{3}\pi r^3} = 3r^{-1}$$

$$= 3 \times 2/20 \times 10^{-6}\,\text{m} = 3.0 \times 10^5\,\text{m}^{-1}$$

Thus, the ratio of these two surface-to-volume ratios is

$$E.\ coli/\text{eukaryotic} = 5 \times 10^6\,\text{m}^{-1}/3.0 \times 10^5\,\text{m}^{-1} = \textbf{17}$$

Since cells must take in all nutrients through their surfaces, the *E. coli* cell can absorb nutrients 17 times faster per unit volume. Thus, an *E. coli* cell can have a 17 times greater metabolism per unit volume than the eukaryotic cell, all else being equal.

A single microvillus adds the volume

$$\pi \times (0.1 \times 10^{-6}\ \text{m}/2)^2 \times 1 \times 10^{-6}\,\text{m} = 7.85 \times 10^{-21}\,\text{m}^3$$

and the surface area

$$\pi \times 0.1 \times 10^{-6}\,\text{m} \times 1 \times 10^{-6}\,\text{m} = 3.14 \times 10^{-13}\,\text{m}^2 \ [1]$$

to the brush border cell (the top of the cylinder is not added surface area since the cell has this surface area without the microvilli). The area on the eukaryotic cell that is covered with microvilli is

$$0.20 \times 4\pi \times (20 \times 10^{-6}/2)^2 = 2.5 \times 10^{-10} \text{ m}^2$$

There is one microvillus per $(0.2 \times 10^{-6} \text{ m})^2 = 4 \times 10^{-14} \text{ m}^2$

Number of microvilli = $2.5 \times 10^{-10} \text{ m}^2/4 \times 10^{-14} \text{ m}^2 = 6250$

Volume of cell with microvilli $= \frac{4}{3}\pi \times (20 \times 10^{-6} \text{ m}/2)^3 + 6250 \times 7.85 \times 10^{-21} \text{ m}^3$

$$= 4.19 \times 10^{-15} \text{ m}^3 + 4.91 \times 10^{-17} \text{ m}^3 = 4.23 \times 10^{-15} \text{ m}^3$$

Area of the brush border cell = $4\pi \times (20 \times 10^{-6} \text{ m}/2)^2 + 6250 \times 3.14 \times 10^{-13} \text{ m}^2$

$$= 1.26 \times 10^{-9} \text{ m}^2 + 1.96 \times 10^{-9} \text{ m}^2 = 3.16 \times 10^{-9} \text{ m}^2$$

Surface-to-volume ratio of cell with microvilli = $3.16 \times 10^{-9} \text{ m}^2/4.23 \times 10^{-15} \text{ m}^3 = 7.47 \times 10^5 \text{ m}^{-1}$

Thus, the microvilli have increased the surface-to-volume ratio of the brush border cell by a factor of $7.47 \times 10^5/3.0 \times 10^5 = \mathbf{2.49.}$

4. Volume of *E. coli* cell = $\pi (1 \times 10^{-6} \text{ m}/2) \times 2 \times 10^{-6} \text{ m} \times 10^3 \text{ L} \bullet \text{m}^{-3} = 1.57 \times 10^{-15} \text{ L}$

   Number of moles of the protein in an *E. coli* = 2 molecules$/6.02 \times 10^{23}$ molecules$\bullet$mol^{-1}

$$= 3.32 \times 10^{-24} \text{ mol}$$

   Concentration of the protein = 3.32×10^{-24} mol $/1.57 \times 10^{-15}$ L = $\mathbf{2.11 \times 10^{-9}}$ *M*

   1 m*M* concentration contains $6.02 \times 10^{23} \times 10^{-3}$ molecules$\bullet$L^{-1}= 6.02×10^{20} molecules$\bullet$L^{-1}

   Number of molecules in an *E. coli* cell = 1.57×10^{-15} L $\times 6.02 \times 10^{20}$ molecules$\bullet$L^{-1}

$$= \mathbf{9.45 \times 10^5 \text{ molecules of glucose}}$$

5. Volume of *E. coli* cell = $\pi \times (1 \times 10^{-6}/2)^2 \times 2 \times 10^{-6} \text{ m}^3 = 1.57 \times 10^{-18} \text{ m}^3$

   Volume of DNA = $\pi \times (20 \text{ Å} \times 10^{-10} \text{ m} \bullet \text{Å}^{-1}/2)^2 \times 1.4 \text{ mm} \times 10^{-3} \text{ m/mm} = 4.40 \times 10^{-21} \text{ m}^3$

   Fraction of volume of an *E. coli* cell occupied by DNA = $4.40 \times 10^{-21} \text{ m}^3/1.57 \times 10^{-18}$

$$= \mathbf{2.80 \times 10^{-3}}$$

$$\text{Volume of human cell} = \frac{4}{3}\pi \times (20 \times 10^{-6}/2)^3 \text{ m}^3 = 4.19 \times 10^{-15} \text{ m}^3$$

Volume of human DNA = 700 × volume of *E. coli* DNA = 700 × 4.40 × 10^{-21} m^3

$$= 3.08 \times 10^{-18} \text{ m}^3$$

Fraction of volume of human cell occupied by DNA = 3.08 × 10^{-18} m^3/4.19 × 10^{-15} m^3

$$= 7.35 \times 10^{-4}$$

6. Since it is likely that any life forms on the planet will be microscopic, they will have to be detected by chemical means. Such life forms may not have macromolecules that resemble those of terrestrial life forms but they will most probably have some sort of a carbon based metabolism. Thus, if the life forms were supplied with radioactively labeled nutrients, their ability to incorporate the label into new compounds would indicate their existence. [Such experiments were carried out on Mars by the Viking landers in 1976. The results were negative. See Horowitz, H. N., The Search for Life on Mars, *Sci. Am.* **237**(5): 52-61 (1977).]

7. With the earth so cold and dark, those eukaryotic species unable to withstand the cold, particularly those from the tropics, would rapidly die out. As time passed, surviving plants would be unable to photosynthesize so that they and the other surviving eukaryotes would eventually starve. Once the nuclear winter had passed, the destruction of almost all ecosystems would further aid in the demise of eukaryotes. Although prokaryotes would also be badly affected by a nuclear winter, their capacity to remain dormant and even frozen for nearly indefinite time periods would help them survive this difficult period. Their ability to live in environments hostile to eukaryotes together with their ability to rapidly adapt to new environments would further aid their survival after the nuclear winter.

8. The purple and green photosynthetic bacteria are obligate anaerobes that require the presence of CO_2 and H_2S to carry out photosynthesis. It is therefore likely that these compounds, but not O_2, were abundant in the earth's atmosphere when these organisms first arose.

Chapter 2
AQUEOUS SOLUTIONS

1.

2. (a) Chloroform is a polar molecule whereas CCl_4 is not. Alignment of the $CHCl_3$ dipole in an electric field will oppose the field thereby decreasing the effective field strength.

(b) Ethanol has a larger nonpolar group than does methanol so that the effect of the polar part of the molecule on the dielectric constant is less in ethanol in comparison to that of methanol.

(c) Formamide has a much larger dipole moment than does acetone so that its dielectric constant is larger for the same reason as given in Part (a). In addition, formamide molecules can hydrogen bond to one another whereas acetone molecules cannot. This reduces the thermally induced reorientations of formamide molecules in an electric field relative to those of acetone.

3. The nonpolar tails of amphiphiles are soluble in nonpolar solvents whereas the polar head groups are not. Hence, the heads aggregate so as to be out of contact with the nonpolar solvent whereas the tails are solvated by it. The forces that stabilize the inverted micelle are therefore the opposite of the hydrophobic and hydrophilic forces stabilizing micelles in aqueous solutions.

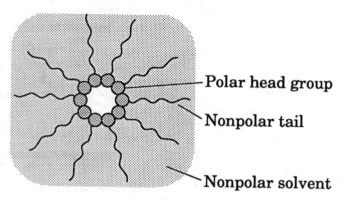

4. Hydrophobic forces tend to expel the nonpolar portions of surfactants from the aqueous phase. Hence, surfactants gather at phase boundaries with their nonpolar groups projecting out of the aqueous phase and into the (presumably) less polar phase. The presence of these nonpolar entities disrupts the hydrogen bonded network of surface water molecules which is responsible for water's large surface tension.

The nonpolar tails of surfactants are relatively soluble in oily substances and oily dirt. They coat these substances as shown

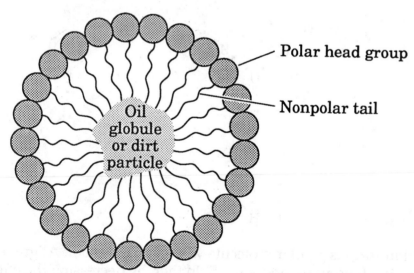

to form a micellelike assembly that has a polar surface. This is soluble (dispersible) in water whereas the oily particle itself, being hydrophobic, is not.

5. Hydrogen bonding forces, being electrostatic in nature, increase, according to equation [2.1], as the dielectric constant of the medium decreases. Hydrophobic forces decrease with decreasing solvent dielectric constant. Substances of low dielectric constant tend to be relatively nonpolar. Nonpolar solvent molecules associate with nonpolar solutes

more than with H_2O molecules. Therefore, substances of low dielectric constant decrease the hydrophobic interactions of nonpolar solutes with solvent water.

6. The mobility of K^+ is 76.1×10^{-5} $cm^2 \bullet V^{-1}$ s^{-1}. Hence, in a field of 100 $V \bullet cm^{-1}$, K^+ moves at a velocity of $76.1 \times 10^{-5} \times 100 = 76.1 \times 10^{-3}$ $cm \bullet s^{-1}$. K^+ will therefore traverse 1 cm in $1 cm/76.1 \times 10^{-3}$ $cm \bullet s^{-1} = $ **13.1 s.**

   Similarly, H^+, with an ionic mobility of 362.4×10^{-5} $cm^2 \bullet V^{-1}$ s^{-1}, will traverse 1 cm in 1 $cm/(100$ $V \bullet cm^{-1} \times 362.4 \times 10^{-5}$ $cm^2 \bullet V^{-1}$ $s^{-1}) = $ **2.759 s.**

7. In NaCl, the Na^+ as well as the Cl^- ions are rigidly held in a crystalline lattice and hence are immobile in an electric field. The same is true of entire water molecules in a crystal of ice. However, since water molecules in ice are oriented with their O—H bonds pointing at neighboring water molecules (Figure 2-3), the protons can easily jump from molecule to molecule as is indicated in Figure 2-8. Consequently, the apparent mobility of protons in ice is comparible to that in liquid water because the mechanism of proton migration is similar in both phases.

8. All the substances in this question are strong acids or strong bases and hence are fully ionized in aqueous solutions.

   (a) $[H^+] = 0.1M$ so that $pH = -\log [H^+] = $ **1**

   (b) $[OH^-] = 0.1M$

   $[H^+] = K_w/[OH^-] = 10^{-14}/0.1 = 10^{-13}M$

   so that $pH = $ **13**

   (c) If we ignore the ionization of water, $[H^+] = 3 \times 10^{-5}M$. Since this value is a factor of 333 greater than the value of $[H^+]$ for pure water, the ionization of water can be realistically ignored. Therefore, $pH = $ **4.52.**

   (d) If we ignore the ionization of water, $[H^+] = 5 \times 10^{-10}M$. But $[H^+]$ for H_2O is 200 times greater than this value. Hence, the presence of $HClO_4$ at such a low concentration can be ignored so that $pH = $ **7**, that of pure water.

   (e) $[OH^-] = 2 \times 10^{-8}M$ which is close to that in pure water. Hence the ionization of water cannot be ignored.

   $$[H^+] [OH^-] = 10^{-14}$$

   Charge balance is described by $[H^+] + [Na^+] = [OH^-]$ so that

   $$[H^+] ([H^+] + [Na^+]) = 10^{-14}$$

$$[H^+]^2 + 2 \times 10^{-8}[H^+] - 10^{-14} = 0$$

The quadratic equation yields

$$[H^+] = \frac{-2 \times 10^{-8} \pm \sqrt{(2 \times 10^{-8})^2 + 4 \times 10^{-14}}}{2}$$

$$= \frac{-2 \times 10^{-8} \pm 2 \times 10^{-7}}{2}$$

Choosing the positive root as this yields the only positive and hence physically realistic concentration,

$$[H^+] = 0.9 \times 10^{-7}M \text{ so that pH} = \textbf{7.05}.$$

9. At pH 7, $[H^+] = 10^{-7}$ moles•L^{-1} so that there are $6 \times 10^{23} \times 10^{-7} = 6 \times 10^{16}$ protons•$L^{-1} = 6 \times 10^{13}$ protons•cm^{-3}. The volume of 1 $(\mu m)^3$ is equivalent to $(10^{-4} cm)^3 = 10^{-12} cm^3$ so that this volume contains 6×10^{13} protons•$cm^{-3} \times 10^{-12} cm^3 = \textbf{60 protons}$. Clearly, at this pH individual acid–base groups rarely encounter a hydronium ion in solution.

10. (a)

$$\frac{[H^+][Ac^-]}{[HAc]} = 1.74 \times 10^{-5} M$$

$$x = [H^+] = [Ac^-]$$

$$\frac{x^2}{0.01 - x} = 1.74 \times 10^{-5}$$

$$x^2 + 1.74 \times 10^{-5}x - 1.74 \times 10^{-7} = 0$$

$$x = \frac{-1.74 \times 10^{-5} \pm \sqrt{(1.74 \times 10^{-5})^2 + 4 \times 1.74 \times 10^{-7}}}{2}$$

$$= \frac{-1.74 \times 10^{-5} \pm 8.34 \times 10^{-4}}{2}$$

$$= \textbf{4.08} \times \textbf{10}^{-4}\textbf{M} = [H^+] = [Ac^-]$$

$$[HAc] = 10^{-2} - 4.08 \times 10^{-4} = \textbf{9.59} \times \textbf{10}^{-3}\textbf{M}$$

$$[OH^-] = 10^{-14}/[H^+] = \textbf{2.45} \times \textbf{10}^{-11}\textbf{M}$$

$$pH = -\log[H^+] = \textbf{3.39}$$

(b)
$$\frac{[H^+][NH_3]}{[NH_4^+]} = 5.62 \times 10^{-10} \; M$$

$$x = [H^+] = [NH_3]$$

$$\frac{x^2}{0.25 - x} = 5.62 \times 10^{-10}$$

$$x^2 + 5.62 \times 10^{-10} x - 1.41 \times 10^{-10} = 0$$

$$x = \frac{-5.62 \times 10^{-10} \pm \sqrt{(5.62 \times 10^{-10})^2 + 4 \times 1.41 \times 10^{-10}}}{2}$$

$$x = 1.19 \times 10^{-5} = [H^+] = [NH_3]$$

$$[NH_4^+] = 0.25 - 1.19 \times 10^{-5} = \mathbf{0.25M}$$

$$[OH^-] = 10^{-14}/[H^+] = \mathbf{8.4 \times 10^{-10} M}$$

$$[Cl^-] = \mathbf{0.25M}$$

$$pH = \mathbf{4.92}$$

(c) According to equation [2.9],

$$pH = pK + \log\left(\frac{x}{c_o - x}\right)$$

where $c_o = 0.05 + 0.10 = 0.15M$

$$x = 0.10M$$

$$pH = 4.76 + \log\frac{0.10}{0.05} = 5.06$$

$$[H^+] = \mathbf{8.71 \times 10^{-6} M}$$

$$[OH^-] = 10^{-14}/8.70 \times 10^{-6} = \mathbf{1.15 \times 10^{-9} M}$$

$$[Ac^-] = \mathbf{0.10M} \qquad \text{(unchanged from its initial value)}$$

$$[HAc] = \mathbf{0.05M} \qquad \text{(unchanged from its initial value)}$$

$$[Na^+] = \mathbf{0.10M}$$

(d)
$$c_o = 0.20 + 0.05 = 0.25M$$

$$x = 0.05M$$

$$pH = pK + \log\frac{0.05}{0.20} = 9.24 + \log 0.25 = 8.64$$

$[H^+] = 2.30 \times 10^{-9}M$

$[OH^-] = 10^{-14}/2.30 \times 10^{-9} = 4.34 \times 10^{-6}M$

$[B(OH)_3] = 0.20M$ (unchanged from its initial value)

$[B(OH)_4^-] = 0.05M$ (unchanged from its initial value)

$[Na^+] = 0.05M$

11. (a) The endpoint of an acetic acid titration lies around pH 9 (depending on the total acetic acid + acetate concentration) so that phenolphthalein is an effective indicator for its titration; that is, it changes color over the pH range containing the titration end point.

(b) The endpoint of NH_4^+ titration lies at a pH above the color change range of phenolphthalein.

(c) Only the second equivalence point of H_3PO_4 ($H_2PO_4^-$ $HPO_4^{2-} + H^+$) lies in the proper pH range; the first ionization lies below it and the third lies above it.

12. Since the three pK's of phosphoric acid are well separated, equation [2.9] can be used:

$$pH = pK_2 + \log\left(\frac{x}{c_o - x}\right)$$

where $x = 0.12M$ and $c_o = 0.12 + 0.08 = 0.20M$.

$$pH = 7.20 + \log\frac{0.12}{0.08} = 7.38$$

$[H^+] = 4.17 \times 10^{-8}M$

$[OH^-] = 10^{-14}/4.17 \times 10^{-8} = 2.40 \times 10^{-7}M$

$[H_2PO_4^-] = 0.08M$ (its initial value)

$[HPO_4^{2-}] = 0.12M$ (its initial value)

$[H_3PO_4] = [H^+][H_2PO_4^-]/K_1 = 4.17 \times 10^{-8} \times 0.08/7.08 \times 10^{-3}$

$$= 4.71 \times 10^{-7}M$$

$[PO_4^{3-}] = K_3[HPO_4^{2-}]/[H^+] = 4.17 \times 10^{-13} \times 0.12/4.17 \times 10^{-8}$

$$= 1.20 \times 10^{-6}M$$

These latter two calculations prove the initial assumption that $[H_3PO_4]$ and $[PO_4^{3-}]$ are negligible compared to $[H_2PO_4^-]$ and $[HPO_4^{2-}]$.

$$[K^+] = 0.08 + 2 \times 0.12 = \mathbf{0.32M}$$

13.
$$H_2O + CO_2 \rightleftharpoons H_2CO_3$$

$$H_2CO_3 \rightleftharpoons H^+ + HCO_3^- \qquad K = \frac{[H^+][HCO_3^-]}{[H_2CO_3]} = 4.47 \times 10^{-7} M$$

Since the second ionization of H_2CO_3 is well separated from the first:

$$[H^+] = [HCO_3^-] = x$$

$$\frac{x^2}{c_o - x} = \frac{x^2}{1.0 \times 10^{-5} - x} = 4.47 \times 10^{-7}$$

$$x^2 + 4.47 \times 10^{-7}x - 4.47 \times 10^{-12} = 0$$

$$x = \frac{-4.47 \times 10^{-7} \pm \sqrt{(4.47 \times 10^{-7})^2 + 4 \times 4.47 \times 10^{-12}}}{2}$$

$$= \frac{-4.47 \times 10^{-7} \pm \sqrt{1.81 \times 10^{-11}}}{2} = 1.90 \times 10^{-6} M$$

since only the positive root yields a positive concentration.

$$pH = -\log [H^+] = \mathbf{5.72}$$

14. Using equation [2.9]:

$$pH = pK + \log\left(\frac{x}{c_o - x}\right)$$

$$5.0 = 4.76 + \log\left(\frac{x}{0.20 - x}\right) \quad \text{where } x = [Ac^-].$$

$$\frac{x}{0.20 - x} = 10^{5.00 - 4.76} = 10^{0.24} = 1.74$$

$$x = 0.35 - 1.74x$$

$$x = \frac{0.35}{2.74} = \mathbf{0.13M} = [Ac^-]$$

$$[HAc] = 0.20 - 0.13 = \mathbf{0.07M}$$

15. At pH 7, only citrate^{2-} and citrate^{3-} have significant concentrations as only the pK of this ionization of citric acid is near pH 7. Using equation [2.9], initially

$$pH = pK + \log\left(\frac{x}{c_o - x}\right)$$

where $x = $ [citrate^{3-}] and $c_o = 0.120M$.

$$7.00 - 6.40 = \log\frac{x}{0.120 - x}$$

$$\frac{x}{0.120 - x} = 10^{0.60} = 3.98$$

$$x = 0.478 - 3.98x$$

$$x = \frac{0.478}{4.98} = 9.59 \times 10^{-2}M = \text{[citrate}^{3-}\text{]}$$

$$\text{[citrate}^{2-}\text{]} = 0.120 - 9.59 \times 10^{-2} = 2.41 \times 10^{-2}M$$

The reaction increases [citrate^{2-}] by 0.20×10^{-3} equivalents/10×10^{-3} L $= 0.02M$ so that the final concentrations are:

$$\text{[citrate}^{2-}\text{]} = 2.41 \times 10^{-2} + 2.0 \times 10^{-2} = 4.41 \times 10^{-2}M$$

$$\text{[citrate}^{3-}\text{]} = 9.59 \times 10^{-2} - 2.0 \times 10^{-2} = 7.59 \times 10^{-2}M$$

$$pH_{final} = pK + \log\frac{7.59 \times 10^{-2}}{4.41 \times 10^{-2}}$$

$$pH_{final} = 6.40 + \log 1.72 = \textbf{6.64}$$

In the absence of any buffer, $[H^+]_{final} = 0.02$ so that $pH_{final} = \textbf{1.70}$.

16.

$$\beta = \frac{\Delta x}{\Delta pH} \approx \frac{dx}{dpH} = \frac{1}{dpH/dx}$$

where x is the incremental amount of base added in equivalents. Hence, differentiating equation [2.9] and using the relationship $\ln q = 2.303 \log q$,

$$\beta = \frac{2.303 \, x \, (c_o - x)}{c_o} = \frac{2.303 \, [A^-][HA]}{[A^-] + [HA]}$$

The point of maximum buffer capacity is determined by differentiating this equation and setting the result equal to zero:

$$\frac{d\beta}{dx} = 2.303\,(c_o - 2x)/c_o = 0$$

Buffer capacity is therefore maximal at

$$x = c_o/2$$

which is the midpoint of the titration curve where $[A^-] = [HA]$ and $pH = pK$.

17. From equation [2.10],

$$K_1 = K_A + K_B$$

and

$$K_2 = \frac{1}{1/K_C + 1/K_D}$$

But since both oxalic and succinic acids are symmetrical, $K_A = K_B$ and $K_C = K_D$. Therefore:

$$K_1 = 2K_A$$

$$K_A = K_1/2$$

and

$$K_2 = K_C/2$$

$$K_C = 2K_2$$

For oxalic acid: $K_1 = 5.37 \times 10^{-2}$ and $K_2 = 5.37 \times 10^{-5}$

Hence,

$$K_A = K_B = 5.37 \times 10^{-2}/2 = \mathbf{2.68 \times 10^{-2}}$$

and

$$K_C = K_D = 2 \times 5.37 \times 10^{-5} = \mathbf{1.07 \times 10^{-4}}$$

For succinic acid: $K_1 = 6.17 \times 10^{-5}$ and $K_2 = 2.29 \times 10^{-6}$

$$K_A = K_B = 6.17 \times 10^{-5}/2 = \mathbf{3.08 \times 10^{-5}}$$

and

$$K_C = K_D = 2 \times 2.29 \times 10^{-6} = \mathbf{4.58 \times 10^{-6}}$$

Chapter 3
THERMODYNAMIC PRINCIPLES: A REVIEW

1. Both are statements of the drive towards randomness of all spontaneous processes.

2. Since climbing the stairs will produce little frictional heat,

$$\Delta U = w = Fh = mgh$$

where h is the height climbed. But,

$$\Delta U = 500 \text{ Cal} \times 4184 \text{ J/Cal} \times 0.20 = 418,400 \text{ J}$$

$$h = \frac{\Delta U}{mg} = \frac{418,400 \text{ kg} \bullet \text{m}^2 \bullet \text{s}^{-2}}{75 \text{ kg} \times 9.8 \text{ m} \bullet \text{s}^{-2}} = 569 \text{ m}$$

Therefore,

$$\text{Flights of stairs} = \frac{569 \text{ m}}{4 \text{ m/flight}} = \textbf{142 flights}$$

3. There are only a limited number of configurations that constitute an acceptably parked car in a small space. In contrast, there are a much larger number of configurations that constitute a car which has been pulled out of a small parking place. Hence, there is a greater entropy increase associated with pulling the car out of the parking space than there is associated with parking it. Since there is no enthalpy change associated with the position of the car (at least in a flat parking lot), pulling a car out of a parking space is a

more spontaneous process (lesser ΔG) than is parking the car.

4. The phrase has 18 characters. There are $(46)^{18}$ different 18-character phrases possible. Half would have to be typed, on average, before the desired phrase was generated. This would take one million monkeys, each typing one key stroke per second, $(46)^{18}/2 \times 10^6 =$ **4.25×10^{23} s $= 1.35 \times 10^{16}$ years** (well over the 2.0×10^{10} year estimated age of the universe). This calculation indicates that the probability of randomly generating even a relatively simple ordered system from disorder is essentially nil.

5. Since q is positive for heat absorbed, the entropy change associated with loosing an amount ot heat, q, at T_1 and absorbing it at T_2 is :

$$\Delta S = \frac{-q}{T_1} + \frac{q}{T_2} = \frac{(T_1 - T_2)q}{T_1 T_2}$$

For $T_1 > T_2$, $\Delta S > 0$, a spontaneous process.

For $T_1 < T_2$, $\Delta S < 0$, a nonspontaneous process.

6. Each CO molecule in the crystal can have two orientations: the C to the left or the C to the right. Consequently, there are only two ways of lining up all of the CO molecules head-to-tail: all C's to the left or all C's to the right. The total number of ways of lining up 10^{23} CO molecules is $2^{10^{23}}$. The probability, W, of randomly obtaining all head-to-tail molecules is therefore $2/2^{10^{23}}$ so that

$$S = k_B \ln W = 1.3807 \times 10^{-23} \text{ J} \cdot \text{K}^{-1} [\ln 2 - 10^{23} \ln 2] = -0.96 \text{ J} \cdot \text{K}^{-1} = \mathbf{-0.96 \text{ J} \cdot \text{K}^{-1}}$$

7. The answer here is similar to that of Problem 5. The perpetual motion machine of the second kind proposed transfers heat, q, at the temperature of the sea, say T_1, to boiling water at temperature T_2 where $T_2 > T_1$. For such a process

$$\Delta S = \frac{-q}{T_1} + \frac{q}{T_2} = \frac{(T_1 - T_2)q}{T_1 T_2} < 0$$

where ΔS refers to the entire system, that is the ship and the ocean. Since the 2nd law of thermodynamics states that $\Delta S > 0$ for real processes, the proposed ship's engine cannot operate as advertised.

8. $$\Delta G^\circ = \Sigma G_f^\circ(\text{products}) - \Sigma G_f^\circ(\text{reactants})$$

(a) $\Delta G^\circ = 6(-386.2) + 6(-237.2) - (-917.2) - 6(0) = \mathbf{-2823.2 \text{ kJ} \cdot \text{mol}^{-1}}$

(b) $\Delta G^\circ = 2(-181.5) + 2(-386.2) - (-917.2) = \mathbf{-218.2 \text{ kJ} \cdot \text{mol}^{-1}}$

(c) $\Delta G^\circ = 2(-516.6) + 2(0) - (-917.2) = \mathbf{-116.0 \text{ kJ} \cdot \text{mol}^{-1}}$

9. (a) For the reaction

$$\text{Protein }(denatured) \rightleftharpoons \text{protein }(native)$$

$$K = \frac{[\text{protein}(native)]}{[\text{protein}(denatured)]}$$

T(°C)	T(K)	1/T	[Protein (denatured)]	[Protein (native)]	K	ln K
50	323	3.10×10^{-3}	$5.1 \times 10^{-6} M$	$2.0 \times 10^{-3} M$	392	5.97
100	373	2.68×10^{-3}	$2.8 \times 10^{-4} M$	$1.7 \times 10^{-3} M$	6.1	1.80

$$\ln K = \frac{-\Delta H^\circ}{R}\left(\frac{1}{T}\right) + \frac{\Delta S^\circ}{R}$$

This is a linear equation of form $y = mx + b$ where $y = \ln K$, $m = -\Delta H^\circ/R$, $x = 1/T$, and $b = \Delta S^\circ/R$. We have two different conditions: $y_1 = mx_1 + b$ and $y_2 = mx_2 + b$. Subtracting them

$$y_1 - y_2 = m(x_1 - x_2)$$

$$m = \frac{y_1 - y_2}{x_1 - x_2} = \frac{-\Delta H^\circ}{R}$$

$$\Delta H^\circ = -8.314 \ \text{J} \bullet \text{K}^{-1} \text{mol}^{-1} \times \left\{ \frac{5.97 - 1.80}{(3.10 - 2.68) \times 10^{-3} \ \text{K}^{-1}} \right\} = -82.5 \times 10^3 \ \text{J} \bullet \text{mol}^{-1}$$

Adding the two equations:

$$y_1 + y_2 = m(x_1 + x_2) + 2b$$

$$b = \frac{1}{2}\left\{ y_1 + y_2 - m(x_1 + x_2) \right\} = \frac{\Delta S^\circ}{R}$$

$$\Delta S^\circ = 8.314 \ \text{J} \bullet \text{K}^{-1} \text{mol}^{-1} \times \frac{1}{2} \times \left\{ 5.97 + 1.80 + \frac{-82.5 \times 10^3}{8.314}\left(3.10 \times 10^{-3} + 2.68 \times 10^{-3} \right) \right\}$$

$$= -206 \ \text{J} \bullet \text{K}^{-1} \ \text{mol}^{-1}.$$

(b) $\Delta G^\circ = \Delta H^\circ - T\Delta S^\circ$ where at 25°C, $T = 298$ K

$$= -82.5 \times 10^3 \ \text{J} \bullet \text{mol}^{-1} - 298 \ \text{K} \times (-206 \ \text{J} \bullet \text{K}^{-1} \ \text{mol}^{-1}) = -21.1 \times 10^3 \ \text{J} \bullet \text{mol}^{-1}$$

Since ΔG° is negative, the process is **spontaneous.**

(c) At the denaturation temperature

$$\Delta G^\circ = \Delta H^\circ - T\Delta S^\circ = 0$$

$$T = \frac{\Delta H^\circ}{\Delta S^\circ} = \frac{-82.5 \times 10^3 \text{ J} \cdot \text{mol}^{-1}}{-206 \text{ J} \cdot \text{K}^{-1} \text{mol}^{-1}} = 400 \text{ K} = \mathbf{127^\circ C}$$

10. (a) In the formation of $H_2O(l)$ from its component elements, the standard states of the elements are the same in the physical chemistry and biochemistry conventions: 25°C, 1 atm, and in their most stable forms. Hence, the standard free energy of formation of $H_2O(l)$ is the same experimentally measurable quantity under both conventions, that is,

$$\Delta G_f^{\circ\prime}[H_2O(l)] = \Delta G_f^\circ[H_2O(l)] = \mathbf{-237.2 \text{ kJ} \cdot \text{mol}^{-1}}$$

(b) $\Delta G^{\circ\prime} = -29.3$ kJ mol^{-1} for the reaction

$$\text{Sucrose} + H_2O \rightarrow \text{glucose} + \text{fructose}$$

where

$$\Delta G^{\circ\prime} = \Delta G_f^{\circ\prime}(\text{glucose}) + \Delta G_f^{\circ\prime}\text{fructose}) - \Delta G_f^{\circ\prime}(\text{sucrose}) - \Delta G_f^{\circ\prime}[H_2O(l)]$$

so that

$$\Delta G_f^{\circ\prime}(\text{sucrose}) = \Delta G_f^{\circ\prime}(\text{glucose}) + \Delta G_f^{\circ\prime}(\text{fructose}) - \Delta G_f^{\circ\prime}[H_2O(l)] - \Delta G^{\circ\prime}$$

Since $\Delta G_f^\circ = \Delta G_f^{\circ\prime}$ for all these quantities,

$$\Delta G_f^{\circ\prime}(\text{sucrose}) = \Delta G_f^\circ(\text{glucose}) + \Delta G_f^\circ(\text{fructose}) - \Delta G_f^\circ[H_2O(l)] - \Delta G^{\circ\prime}$$

$$= -917.2 - 915.4 + 237.2 + 29.3 = \mathbf{-1566.1 \text{ kJ mol}^{-1}}$$

(c) For the reaction

$$\text{Ethyl acetate} + H_2O \rightleftharpoons \text{ethanol} + \left\{ \begin{array}{c} H^+ + OAc^- \\ \Updownarrow \quad K = 10^{-4.76} \\ HOAc \end{array} \right\}$$

where $OAc^- = \text{acetate}^-$, equations [3.20] and [3.21] indicate that

$$\Delta G^{\circ\prime} = \Delta G^\circ - RT \ln [H_2O] - RT \ln (1 + [H^+]/K) + RT \ln [H^+]_o$$

$$= \Delta G^\circ - RT \ln 55.5 - RT \ln (1 + 10^{-7}/10^{-4.76}) + RT \ln 10^{-7}$$

(note that the negative sign preceeding the $RT\ln[H_2O]$ term occurs because water is taken up in the above reaction rather than being produced as in the reaction to which equation [3.20] refers). Hence,

$$\Delta G^{\circ\prime}(\text{ethyl acetate}) = \Delta G^{\circ} - 20.1RT$$

But

$$\Delta G^{\circ} = \Delta G_f^{\circ}(\text{ethanol}) + \Delta G_f^{\circ}(\text{OAc}^-) + \Delta G_f^{\circ}(\text{H}^+) - \Delta G_f^{\circ}(\text{ethyl acetate}) - \Delta G_f^{\circ}(\text{H}_2\text{O})$$

so that,

$$\Delta G_f^{\circ\prime}(\text{ethyl acetate}) = \Delta G_f^{\circ}(\text{ethyl acetate})$$

$$= \Delta G_f^{\circ}(\text{ethanol}) + \Delta G_f^{\circ}(\text{OAc}^-) + \Delta G_f^{\circ}(\text{H}^+) - \Delta G_f^{\circ}(\text{H}_2\text{O}) - \Delta G^{\circ}$$

$$= -181.5 - 369.2 + 0 + 237.2 - \Delta G^{\circ} = -313.5 - \Delta G^{\circ\prime} + 20.1\ RT$$

$$= -313.5 + 19.7 + 49.8 = \mathbf{-244.0\ kJ \cdot mol^{-1}}$$

11. From equation [3.16]:

$$K_{eq}{}' = e^{-\Delta G^{\circ\prime}/RT}$$

(a) Phosphoenolpyruvate:

$$K_{eq}{}' = e^{61.9 \times 1000/8.314 \times 298} = \mathbf{7.1 \times 10^{10}}$$

(b) Pyrophosphate:

$$K_{eq}{}' = e^{33.5 \times 1000/8.314 \times 298} = \mathbf{7.4 \times 10^{5}}$$

(c) Glucose-1-phosphate:

$$K_{eq}{}' = e^{20.9 \times 1000/8.314 \times 298} = \mathbf{4.6 \times 10^{3}}$$

Chapter 4
AMINO ACIDS

1. See Table 4-3.

2. (a)

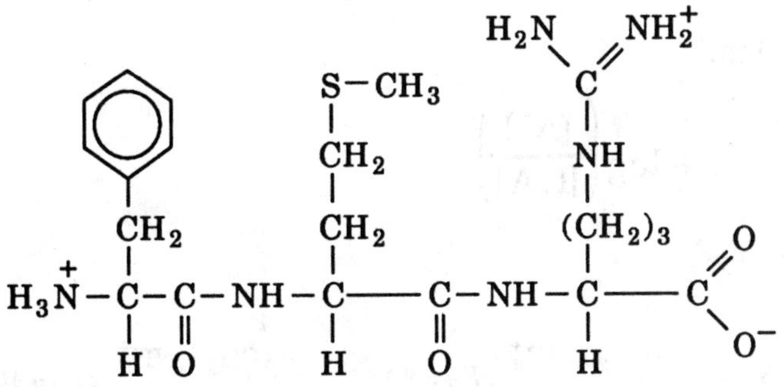

$$H_3\overset{+}{N}-\underset{\underset{H}{|}}{\overset{\overset{CH_2}{|}}{C}}-\underset{\underset{O}{\|}}{C}-NH-\underset{\underset{H}{|}}{\overset{\overset{\overset{\overset{S-CH_3}{|}}{CH_2}}{|}}{\overset{CH_2}{|}}{C}}-\underset{\underset{O}{\|}}{C}-NH-\underset{\underset{H}{|}}{\overset{\overset{\overset{\overset{\overset{H_2N\diagup C\diagdown NH_2^+}{|}}{NH}}{|}}{(CH_2)_3}}{C}}-\underset{\underset{O^-}{}}{\overset{\overset{O}{\|}}{C}}$$

(b)

$$H_3\overset{+}{N}-\underset{}{\overset{\overset{CH_2}{|}}{CH}}-\underset{\underset{O}{\|}}{C}-NH-\underset{}{\overset{\overset{\overset{NH_3^+}{|}}{(CH_2)_4}}{CH}}-\underset{\underset{O}{\|}}{C}-NH-\underset{}{\overset{\overset{\overset{CO_2^-}{|}}{CH_2}}{CH}}-\underset{\underset{O^-}{}}{\overset{\overset{O}{\|}}{C}}$$

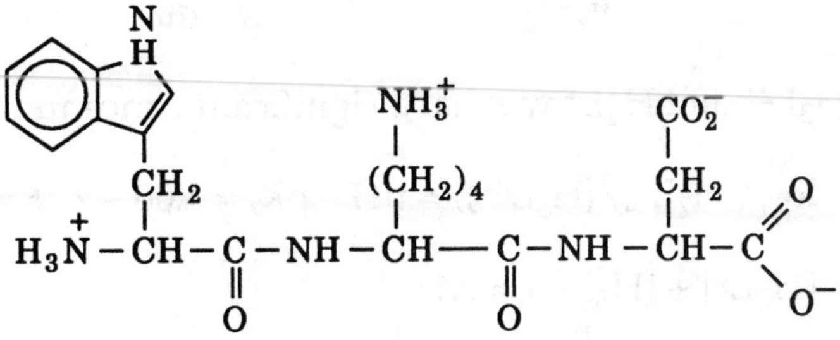

(c)

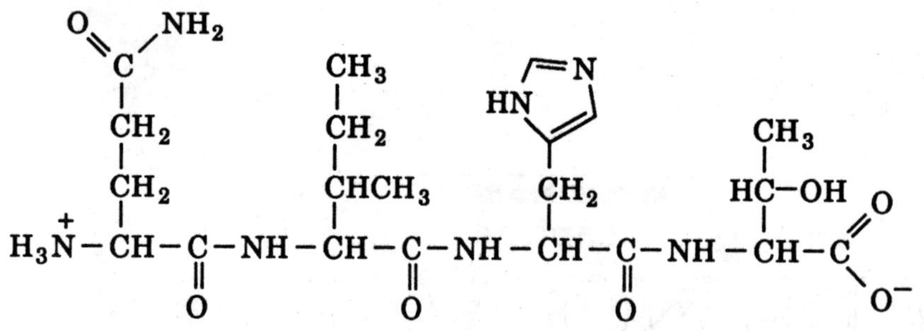

3. The first residue can be one of five residues, the 2nd one of the remaining four, *etc.*

$$N = 5 \times 4 \times 3 \times 2 \times 1 = \mathbf{120}$$

4.

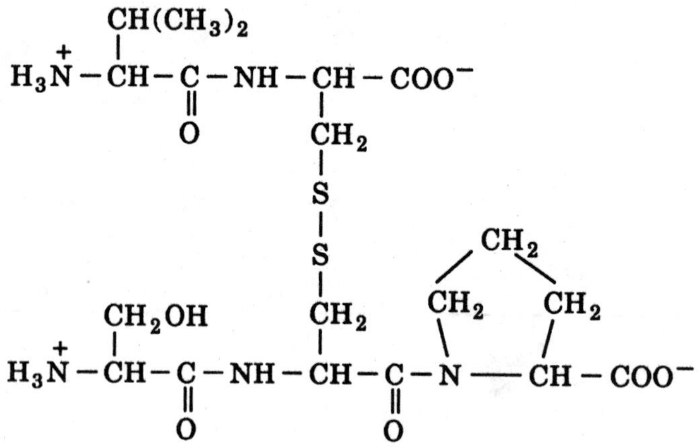

5. Henderson–Hasselbalch:

$$pH = pK + \log\left(\frac{[A^-]}{[HA]}\right)$$

For lysine:

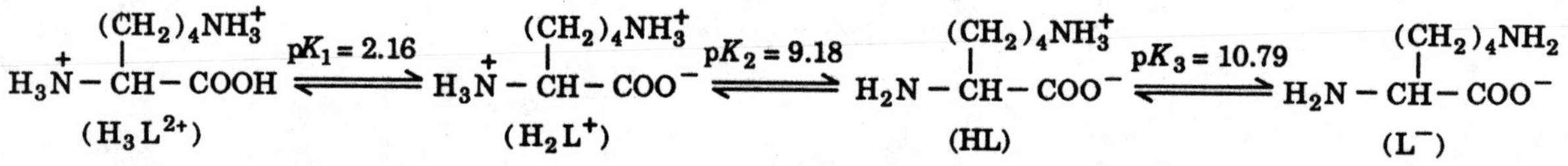

At pH 4, only H_3L^{2+} and H_2L^+ will have significant concentrations:

$$\log([H_2L^+]/[H_3L^{2+}]) = pH - pK_1 = 4.00 - 2.16 = 1.84$$

$$[H_2L^+] + [H_3L^{2+}] = 0.1$$

$$[H_2L^+] = 10^{1.84}[H_3L^{2+}] = 69[H_3L^{2+}]$$

$$(69 + 1)[H_3L^{2+}] = 0.1$$

$$[H_3L^{2+}] = \frac{0.1}{70} = 0.0014M$$

$$[H_2L^+] = 0.1 - 0.0014 = \mathbf{0.0986M}$$

$$[HL] = 10^{[pH - pK_2]} [HL^+] = 10^{4.00 - 9.18} \times 0.0986M = \mathbf{6.5 \times 10^{-7}M}$$

$$[L^-] = 10^{4.00 - 10.79} [HL] = \mathbf{1.1 \times 10^{-13}M}$$

At pH 7, H_2L^+ and HL are the most significant species:

$$\log([HL]/[H_2L^+]) = pH - pK_2 = 7.00 - 9.18 = -2.18$$

$$[HL] + [H_2L^+] = 0.1M$$

$$[HL] = [H_2L^+] 10^{-2.18} = 6.6 \times 10^{-3} [H_2L^+]$$

$$[H_2L^+] (1 + 6.6 \times 10^{-3}) = 0.1M$$

$$[H_2L^+] = \mathbf{0.1M}$$

$$[HL] = 6.6 \times 10^{-3} [H_2L^+] = 6.6 \times 10^{-3} \times 0.1 = \mathbf{6.6 \times 10^{-4}M}$$

$$[H_3L^{2+}] = 10^{pK_1 - pH} [H_2L^+] = 10^{2.16 - 7.00} \times 0.1 = \mathbf{1.4 \times 10^{-6}M}$$

$$[L^-] = 10^{pH - pK_3} [HL] = 10^{10.00 - 10.79} [HL] = 0.16 [HL]$$

At pH 10: First assume only L^- and HL will have significant concentrations:

$$\log([L^-]/[HL]) = pH - pK_3 = 10.00 - 10.79 = -0.79$$

$$[L^-] + [HL] = 0.1M$$

$$[L^-] = [HL] 10^{-0.79} = 0.16[HL]$$

$$(1 + 0.16) [HL] = 0.1$$

$$[HL] = \frac{0.1}{1.16} = 0.086M$$

$$[L^-] = 0.014M$$

$$[H_2L^+] = 10^{pK_2 - pH} [HL] = 10^{9.18 - 10.00} \times 0.086 = 0.03M$$

But $[H_2L^+] \approx [L^-]$ so that the assumption that L^- and HL are the only significant species is incorrect here, and

$$[L^-] + [HL] + [H_2L^+] = 0.1$$

$$\log([HL]/[H_2L^+]) = pH - pK_2 = 10.00 - 9.18$$

$$[H_2L^+] = 10^{pK_2 - pH}[HL] = 10^{9.18 - 10.00}[HL] = 0.15\,[HL]$$

$$\log([L^-]/[HL]) = pH - pK_3 = 10.00 - 10.79$$

$$[L^-] = 10^{pH - pK_3}[HL] = 10^{10.00 - 10.79}[HL] = 0.16\,[HL]$$

$$0.16\,[HL] + [HL] + 0.15\,[HL] = 0.1$$

$$1.31\,[HL] = 0.1$$

$$[HL] = \mathbf{0.076}\boldsymbol{M}$$

$$[H_2L^+] = \mathbf{0.011}\boldsymbol{M}$$

$$[L^-] = \mathbf{0.012}\boldsymbol{M}$$

6.

$$A^- \underset{K_i}{\overset{H^+}{\rightleftharpoons}} HA \underset{K_j}{\overset{H^+}{\rightleftharpoons}} H_2A^+$$

$$pH = pK_i + \log\frac{[A^-]}{[HA]} = pK_j + \log\frac{[HA]}{[H_2A^+]}$$

At the isoelectric point, pH = pI and $[A^-] = [H_2A^+] = x$ so that

$$pK_i = pI - \log\frac{x}{[HA]}$$

$$pK_j = pI - \log\frac{[HA]}{x} = pI + \log\frac{x}{[HA]}$$

Summing these equations:

$$pK_i + pK_j = 2pI$$

so that

$$pI = \frac{1}{2}(pK_i + pK_j)$$

7. For glycine

$$\text{H}_2\text{Gly}^+ \xrightleftharpoons{pK_1 = 2.35} \text{HGly} + \text{H}^+ \xrightleftharpoons{pK_2 = 9.78} \text{Gly}^- + \text{H}^+$$

Mass balance:

$$[\text{Gly}^-] + [\text{HGly}] + [\text{H}_2\text{Gly}^+] = 0.1M$$

Charge balance:

$$[\text{H}^+] + [\text{H}_2\text{Gly}^+] = [\text{Gly}^-] + [\text{OH}^-]$$

Equilibria:

$$\frac{[\text{H}^+][\text{HGly}]}{[\text{H}_2\text{Gly}^+]} = K_1 = 10^{-2.35} = 4.47 \times 10^{-3}$$

$$\frac{[\text{H}^+][\text{Gly}^-]}{[\text{HGly}]} = K_2 = 10^{-9.78} = 1.66 \times 10^{-10}$$

$$[\text{H}^+][\text{OH}^-] = K_w = 10^{-14}$$

If $[\text{H}_2\text{Gly}^+] \gg [\text{H}^+]$ and $[\text{Gly}^-] \gg [\text{OH}^-]$, then the isoionic point is equal to the isoelectric point, pI:

$$\text{pH} = \text{p}I = \frac{1}{2}(\text{p}K_1 + \text{p}K_2) = \frac{1}{2}(2.35 + 9.78) = 6.06$$

Checking the initial assumptions:

$$[\text{H}_2\text{Gly}^+] = \frac{[\text{H}^+][\text{HGly}]}{4.47 \times 10^{-3}}$$

But $[\text{HGly}] \approx 0.1$ so that

$$[\text{H}_2\text{Gly}^+] = \frac{10^{-6.06} \times 0.1}{4.47 \times 10^{-3}} = 1.95 \times 10^{-5}$$

so that

$$\frac{[\text{H}_2\text{Gly}^+]}{[\text{H}^+]} = \frac{1.95 \times 10^{-5}}{10^{-6.06}} = 22$$

Hence, $[\text{H}_2\text{Gly}^+] \gg [\text{H}^+]$ and $[\text{Gly}^-] \gg [\text{OH}^-]$.

8. $$\Delta Z = 13 \; \Delta pH = 13 \; (7.09 - 6.87) = 13 \times 0.22 = 2.86.$$

Therefore, sickle-cell hemoglobin is about 3 ionic charges more negative than normal hemoglobin.

9. (a) chiral
    (b) nonchiral
    (c) chiral (they are usually right handed)
    (d) chiral
    (e) chiral
    (f) prochiral

    (g) nonchiral
    (h) chiral
    (i) prochiral
    (j) prochiral
    (k) chiral
    (l) prochiral

10.

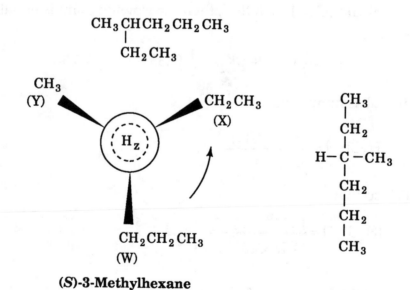

11. (a)

(*S*)-3-Methylhexane

(b)

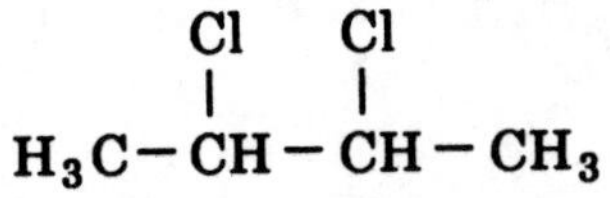

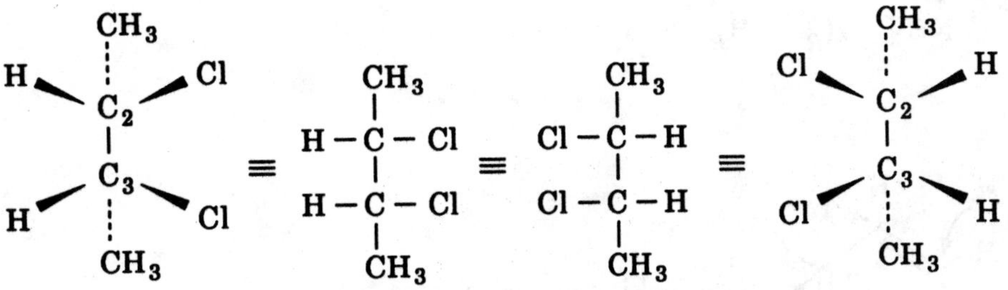

$$Cl \quad Cl$$
$$| \quad\quad |$$
$$H_3C-CH-CH-CH_3$$

$$(2S, 3R) = meso \text{ form} = (2R, 3S)$$

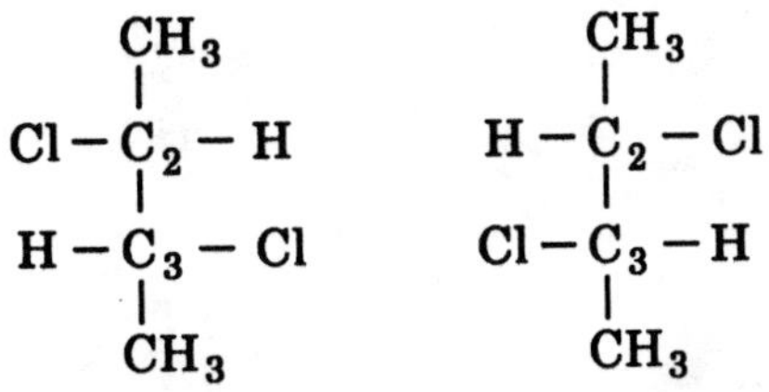

$$(2R, 3R) \quad\quad (2S, 3S)$$

$$[W = Cl, X = -CHClCH_3, Y = CH_3, Z = H]$$

12. (a)   No prochiral centers because the two substituents to the trigonal carbonyl carbon atom are identical.

(b)

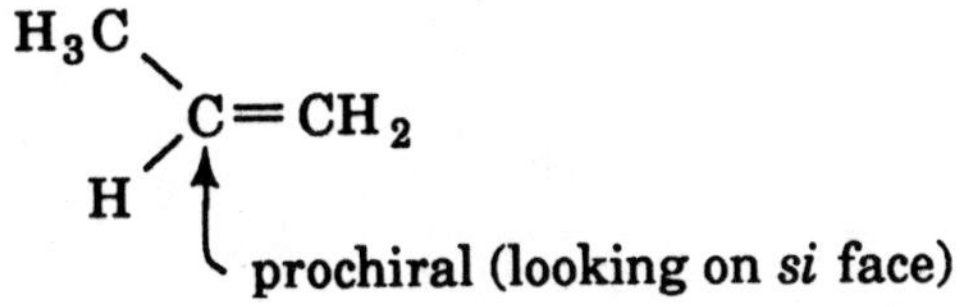

prochiral (looking on *si* face)

(c)

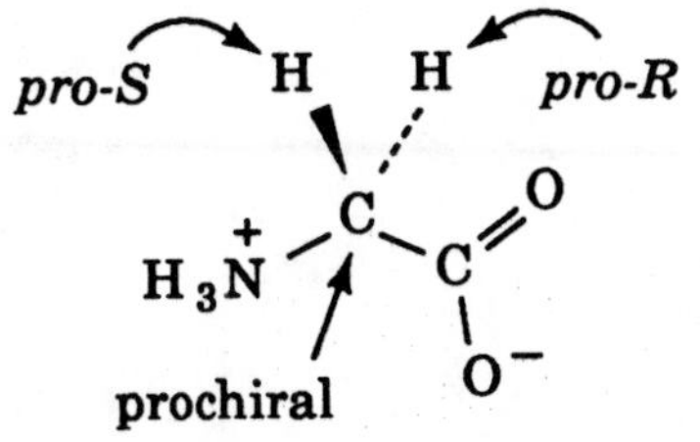

pro-S   H   H   *pro-R*

prochiral

(d)   No prochiral centers.

(e)   In Fischer projection: H_S is *pro-S* and H_R is *pro-R*.

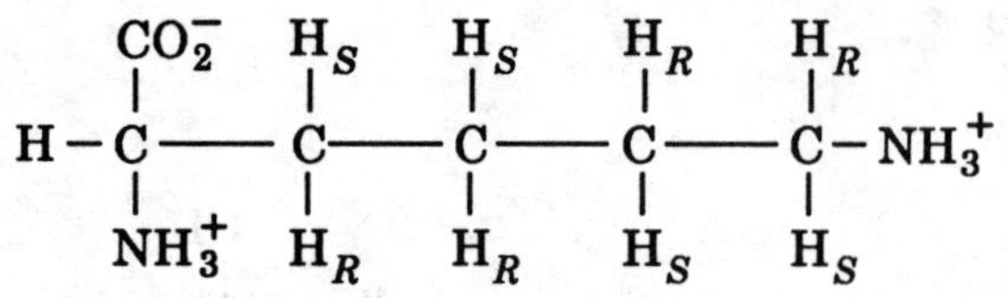

(f)

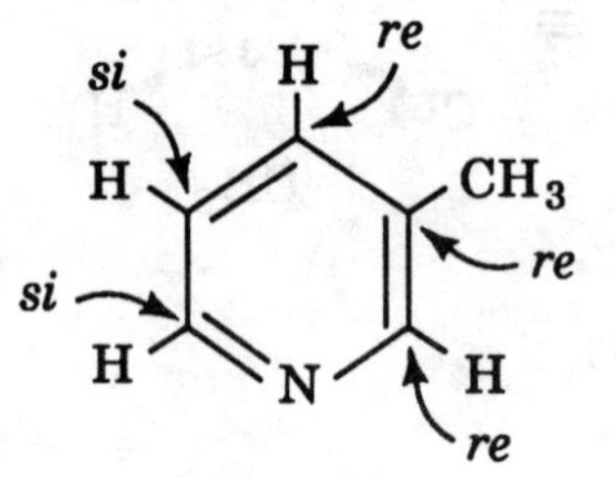

looking at the marked face

Chapter 5

TECHNIQUES OF PROTEIN PURIFICATION

1.

$$I = \frac{1}{2}\sum_i c_i Z_i^2$$

$1M$ NaCl is $1M$ in Na^+ and Cl^-

$$I = \frac{1}{2}(1.0 \times 1^2 + 1.0 \times 1^2) = \mathbf{1.0}$$

$1M$ $(NH_4)_2SO_4$ has $2M$ NH_4^+ + $1M$ SO_4^{2-}

$$I = \frac{1}{2}(2.0 \times 1^2 + 1.0 \times 2^2) = \mathbf{3.0}$$

$1M$ K_3PO_4 has $3M$ K^+ + $1M$ PO_4^{3-}

$$I = \frac{1}{2}(3.0 \times 1^2 + 1.0 \times 3^2) = \mathbf{6.0}$$

Proteins are salted out at high ionic strengths. Hence, **a protein should be most soluble in $1M$ NaCl and least soluble in $1M$ K_3PO_4.**

2. 0.9% NaCl has 9 g NaCl per L of solution. The molecular mass of NaCl is 23 + 35.5 = 58.5 D so that 0.9% NaCl = 9 g•L^{-1}/58.5 g•mol^{-1} = 0.15M.

$$I = \frac{1}{2}(0.15 \times 1^2 + 0.15 \times 1^2) = \mathbf{0.15}$$

3. Dowex 50, as Table 5-2 indicates, has SO_3^- groups that will bind cations. At pH 6, arginine has a charge of +1, aspartate has a charge of –1, histidine (with $pK_R = 6.0$) has a charge of +1/2 and leucine is neutral. Hence, the order of elution is **aspartate, leucine, histidine, arginine.**

4. CM-cellulose has pendent CH_2COO^- groups that bind cations. A protein tends to elute from CM-cellulose faster the lower its isoelectric pH (p*I*) since the higher the p*I*, the more cationic the protein will be at any given pH, and therefore the higher its affinity for the resin. Hence, the order of elution is: **pepsin (pI < 1.0), fibrinogen (5.8), hemoglobin (7.1), ribonuclease (7.8), lysozyme (11.0).**

5. In paper chromatography, the least polar substances move fastest. At pH 4.5, aspartic acid and glutamic acid each have a charge of –1, alanine, phenylalanine, and valine are neutral, and lysine has a charge of +1. The order of decreasing polarity of the neutral amino acids is alanine, valine, and phenylalanine. The order of decreasing polarity of the charged amino acids is aspartatic acid, glutamic acid, and lysine. Hence, the rate of migration (the R_fs) increases as follows:

    aspartatic acid < glutamic acid < lysine < alanine < valine < phenylalanine.

6. Sephadex G-50 is beads of polydextran gel that can fractionate molecules in the molecular mass range 1 to 30 kD with the heaviest eluting first. Hence, according to the molecular masses given in Table 5-5, the order of elution of these proteins from Sephadex G-50 is:

    catalase (222 kD) > concavalin B (42.5 kD) > chymotrypsin α (21.6 kD)

    > myoglobin (16.9 kD) > lipase (6.7 kD).

7. According to Table 5-5, ribonuclease and cytochrome *c* have molecular masses of 12,600 and 13,000 D. Since Sephadex G-50 separates molecules according molecular mass, the unknown protein would have an intermediate molecular mass, about **12,800 D.**

8. According to Table 5-3, Bio-Gel P-30 has an exclusion limit of 40 kD. Hexokinase has a molecular mass larger than this exclusion limit so that it can occupy only the void volume in passing through the column. Thus, the void volume, V_0, of the column is **34 mL,** the elution volume of hexokinase. The volume occupied by gel is, according to equation [5.5],

    $$V_x = V_t - V_0 = 100 - 34 = 66 \text{ mL.}$$

    The relative elution volume of the unknown protein is

    $$V_e/V_0 = 50/34 = 1.47.$$

9. (a) Ala-Phe-Lys and Ala-Ala-Lys differ in their central residue. Since Phe is less polar than Ala, **paper chromatography** can effectively separate these tripeptides.

(b) Since the pI's of lysozyme (pI = 11.0) and ribonuclease (pI = 7.8) differ so much, cation exchange chromatography at a pH between these pI's would separate these proteins. For example, CM-cellulose chromatography at pH 8 should do the job.

(c) Since myoglobin (16.9 kD) and hemoglobin (64.5 kD) differ considerably in molecular mass, gel filtration chromatography on a gel with a fractionation range in the vicinity of these molecular masses, Sephadexes G-50, G-100, G-200, or Bio-Gels P-10, P-30, P-100, or Sepharose 6B (see Table 5-3) will effect their separation.

10. γ–Aminobutyric acid ($H_3N^+CH_2CH_2CH_2COO^-$) can be covalently attached to agarose by BrCN activation. If the covalently attached amino group is not necessary for binding to the receptor, the receptor might be isolated by passing a nerve tissue homogenate over the gel under conditions that the receptor binds γ-aminobutyric acid. After washing the column, the receptor can be removed by elution with a γ-aminobutyric acid solution, or by a solution in which the receptor no longer binds γ-aminobutyric acid (perhaps of different pH or ionic strength).

11. From the data in Table 4-2:

Amino Acid	pK_1	pK_2	pK_R	Ionic Charge at pH 7.5
Arginine	1.8	9.0	12.5	+1
Cysteine	1.9	10.8	8.3	~–0.2
Glutamic acid	2.1	9.5	4.1	–1
Histidine	1.8	9.3	6.0	~0
Leucine	2.3	9.7	–	~0
Serine	2.2	9.2	–	~0

Arginine will migrate rapidly to the cathode. Glutamic acid and cysteine will migrate towards the anode with glutamic acid moving faster than cysteine. Histidine, leucine and serine are essentially electrophoretically immobile at pH 7.5 and hence will not be electrophoretically separated.

12.

Tripeptide	Charge at pH 6.5	Charge at pH 4.5
Asn-Arg-Lys	+2	+2
Asn-Leu-Phe	0	0
Asn-His-Phe	~0.3	+1
Asp-Leu-Phe	–1	–0.8
Val-Leu-Phe	0	0

The order of decreasing polarity at pH 4.5 is:

Asn-Arg-Lys > Asn-His-Phe > Asp-Leu-Phe > Asn-Leu-Phe > Val-Leu-Phe

The fingerprint of these tripeptides will therefore have the appearance:

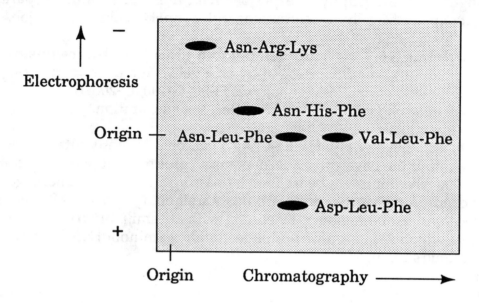

13. The molecular mass in Figure 5-27 that corresponds to a mobility of 0.5 is **28 kD**.

14. Table 5-5 indicates that fibrinogen, with f/f_0 = 2.336, is a highly asymmetric molecule. In its native state it will be more likely to penetrate a given gel pore than a spherical molecule of the same molecular mass. Under gel filtration it will migrate less rapidly than this equivalent spherical molecule and hence appear to have a molecular mass lower than it really has. In contrast, SDS-PAGE denatures fibrinogen so it will migrate at the same rate as almost any other protein of its molecular mass.

15. According to Table 5-1:

Protein	pI
Insulin	5.4
Cytochrome c	10.6
Histone	10.8
Myoglobin	7.0
Ribonuclease	7.8

Since the higher the pI, the more cationic (less anionic) the protein, the proteins will be focused to positions that vary from anode (positive electrode) to cathode (negative electrode) in the order of increasing pI. Thus, from anode to cathode they will take the order **insulin, myoglobin, ribonuclease, cytochrome c and histone.**

16.

$$F_{cent} = m\omega^2 r = ma_{cent}$$

$$a_{cent} = \omega^2 r$$

where r is in cm and ω is in radians$\cdot$s^{-1}.

$$a_{cent} = \left(60{,}000 \text{ rpm} \times 2\pi \text{ radians/revolution} \times \frac{1}{60}\frac{\text{min}}{\text{sec}}\right)^2 \times 6.5 \text{ cm}$$

$$= 2.57 \times 10^8 \text{ cm}\cdot\text{s}^{-2}$$

$$1g = 9.81 \text{ m}\cdot\text{s}^{-2}$$

$$a_{cent} = 2.57 \times 10^8 \text{ cm}\cdot\text{s}^{-2}/9.81 \text{ m}\cdot\text{s}^{-2} \times 100 \text{ cm/m} = \textbf{262{,}000 } \boldsymbol{g}$$

17. From equation [5.15],

$$M = \frac{Nfs}{(1-\bar{V}\rho)}$$

$$M = \frac{6.022 \times 10^{23} \text{mol}^{-1} \times 8.74 \times 10^{-8} \text{g}\cdot\text{s}^{-1} \times 7.35 \times 10^{-13} \text{ s}}{(1 - 0.742 \text{ cm}^3\cdot\text{g}^{-1} \times 0.998 \text{ g}\cdot\text{cm}^{-3})}$$

$$M = \textbf{149{,}000 g/mol (D)}$$

18. Equation [5.15] states that

$$s = \frac{1}{\omega^2}\left(\frac{d\ln r}{dt}\right)$$

which upon integration yields

$$s = \frac{1}{\omega^2}\frac{(\ln r - \ln r_o)}{(t - t_o)} = \frac{1}{\omega^2}\frac{\Delta\ln r}{\Delta t}$$

t (min)	r (cm)	Δt (sec)	$\Delta\ln r$
4	5.944	0	0
6	5.966	120	3.69×10^{-3}
8	5.987	240	7.21×10^{-3}
10	6.009	360	10.88×10^{-3}
12	6.032	480	14.70×10^{-3}

A plot of these data,

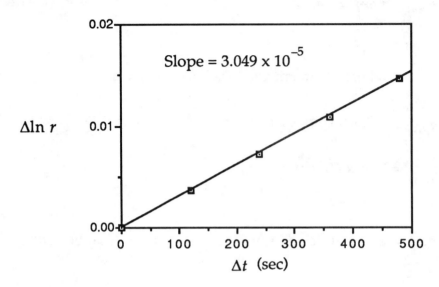

indicates that $\Delta \ln r / \Delta t = 3.049 \times 10^{-5}$. Hence,

$$s = \frac{1}{\omega^2} \frac{\Delta \ln r}{\Delta t} = \frac{3.049 \times 10^{-5}}{\left(\dfrac{35{,}000 \text{ rpm}}{60 \text{ s/min}} \times 2\pi \dfrac{\text{radians}}{\text{revolution}} \right)^2}$$

$$= 2.26 \times 10^{-12} \text{ s} = \mathbf{22.6S}$$

$$M = \frac{Nfs}{(1 - \overline{V}\rho)}$$

$$= \frac{6.022 \times 10^{23} \text{ mol}^{-1} \times 3.72 \times 10^{-8} \text{ g} \cdot \text{s}^{-1} \times 2.26 \times 10^{-12} \text{ s}}{(1 - 0.725 \text{ cm}^3/\text{g} \times 1.030 \text{ g/cm}^3)}$$

$$M = \mathbf{2.00 \times 10^5 \text{ g} \cdot \text{mol}^{-1} \text{ (D)}}$$

Chapter 6

COVALENT STRUCTURES OF PROTEINS

1. (a) Ser-Ala-Phe↓Lys-Pro R_{n-1} = Phe, Trp, Tyr; $R_{n-1} \neq$ Pro

   (b) Thr-Cys-Gly-Met↓Asn R_{n-1} = Met

   (c) Leu-Arg↓Gly↓Asp $R_n \neq$ Arg, Lys, Pro

   (d) Gly↓Phe↓Trp-Pro-Phe-Arg R_n = Ile, Met, Phe, Trp, Tyr, Val; $R_{n-1} \neq$ Pro

   (e) Val-Trp-Lys-Pro-Arg↓Glu R_{n-1} = Arg, Lys; $R_{n-1} \neq$ Pro

2. The protein must have at least three polypeptide chains; one with an N-terminal Ala and two with N-terminal Ser's.

3. The protein C-terminus is (Arg, Lys) with the 3rd residue from the end not Pro. Alternatively, it might consist of two chains, one with a C-terminal Arg and the other with an C-terminal Lys with $R_{n-1} \neq$ Pro for either subunit.

4. (a) All residues but Trp and Gln would be liberated intact (although Ser and Thr would be partially degraded). Trp is destroyed and Gln is liberated as Glu + NH_4^+.

   (b) Cys, Ser and Arg are destroyed. The others are partially deaminated and racemized.

5. Leu is the C-terminus and Pro must be R_{n-2} since carboxypeptidase A only cleaves the

C-terminal residue. Working backwards we have

$$
\begin{array}{r}
\text{Cys-Leu} \\
\text{Pro-Cys-Leu} \\
\text{Ala-Pro} \\
\text{Ser-Ala-Pro} \\
\text{Ser-Ser-Ala} \\
\text{Trp-Ser-Ser} \\
\text{Trp-Ser} \\
\text{Phe-Trp-Ser} \\
\text{Asp-Phe} \\
\text{Lys-Asp-Phe} \\
\text{Ala-Lys} \\
\text{Trp-Ala-Lys} \\
\text{Trp-Ala} \\
\end{array}
$$

Trp-Ala-Lys-Asp-Phe-Trp-Ser-Ser-Ala-Pro-Cys-Leu

6. NCBr cleaves after Met (↓); trypsin cleaves after Arg and Lys (↑). This yields:

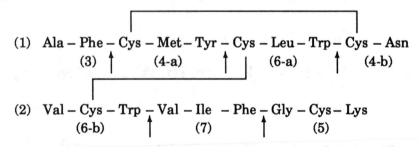

7.

(image shows cross-links connecting cysteine residues)

(1) Ala – Phe – Cys – Met – Tyr – Cys – Leu – Trp – Cys – Asn
       (3) ↑   (4-a)   ↑   (6-a)   ↑   (4-b)

(2) Val – Cys – Trp – Val – Ile – Phe – Gly – Cys – Lys
       (6-b) ↑   (7)   ↑   (5)

(vertical arrows indicate the chymotrypsin cleavage points that yield the indicated polypeptide fragments.)

8. (2) and (3) indicate that Leu and Ser are both N-termini; hence the polypeptide must have 2 chains which are connected by a disulfide bond.

(4) shows that Asp is the C-terminus of one chain; the other C-terminus must be Lys or Arg or be preceeded by Pro.

(5) Ser^N-Arg   (Arg can't be terminal)
(8) Cys-Lys

(6) Met-Asp^C (next to Lys or Arg)
Possibilities:

        (*i*)   Cys-Lys-Met-Asp^C

        (*ii*)   Ser^N-Arg-Met-Asp^C

But (*ii*) is not possible because it lacks a sulfhydryl group to form a disulfide bond with the another chain. Hence, Ser-Arg is N-terminal to either peptide. This yields the possibilities

        (*iii*)   Ser^N-Arg-Cys-Lys-Met-Asp + (7)

or   (*iv*)   Ser^N-Arg-(7)+ Cys-Lys-Met-Asp

But Cys is not N-terminal so (*iv*) must be rejected.

(7 = 12, 13)   Leu^N-Gly-Cys-$\left\{\begin{array}{c}\text{Ile}\\\text{Phe}\end{array}\right\}$-Pro-$\left\{\begin{array}{c}\text{Phe}\\\text{Ile}\end{array}\right\}$

(17)   Leu^N-Gly-Cys-Phe-Pro-Ile^C

Therefore, the primary structure must be

$$\text{Leu} - \text{Gly} - \text{Cys} - \text{Phe} - \text{Pro} - \text{Ile}$$
$$|$$
$$\text{S}$$
$$|$$
$$\text{S}$$
$$|$$
$$\text{Ser} - \text{Arg} - \text{Cys} - \text{Lys} - \text{Met} - \text{Asp}$$

9. From (2) and (3) we see that the peptide has no free terminal ends. Since (1) indicates that the ends aren't chemically blocked by some non-amino acid group, the polypeptide chain must be cyclic.

From (4) and (5) the sequence must be:

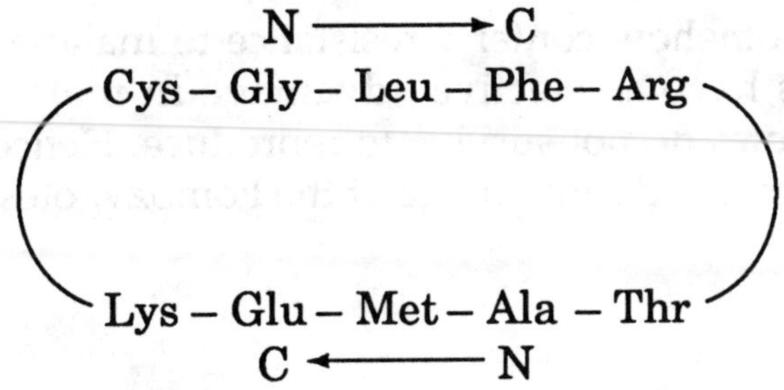

$$N \longrightarrow C$$
$$\text{Cys} - \text{Gly} - \text{Leu} - \text{Phe} - \text{Arg}$$
$$\text{Lys} - \text{Glu} - \text{Met} - \text{Ala} - \text{Thr}$$
$$C \longleftarrow N$$

10. (a)

    (1) Val is C-terminal

    (2) Pro is N-terminal

(3-6)   Pro-Gly-Ala-Arg-(Gly-Lys, Ser-Arg)-Phe-Ile-Val

(b)   The order of the dipeptide Gly-Lys and Ser-Arg is unknown. Derivatizing Lys with citraconic anhydride followed by trypsin hydrolysis and sequencing should yield their order.

11. (a)   The dodo has the fewest differences with the pigeon. It therefore seems most closely related to it.

(b)

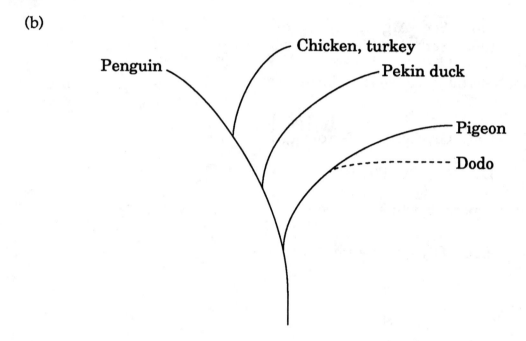

12. Fungi have around 45 to 50 cytochrome *c* amino acid differences with both animals and plants so that fungi are no more closely related to animals than to plants. Hence, the classification of fungi as nongreen plants is about as sensible as classifying them as immobile animals. Fungi should therefore be classified independently from plants and animals.

13. Thalassemia minor must somehow confer a resistance to malaria as does the sickle-cell trait. Thus, heterozygotes have a selective advantage in malarial areas over normal individuals. The homozygotes do not survive to reproduce. Hence, the advantage of the heterozygotes is balanced by the disadvantage of the homozygotes.

14.

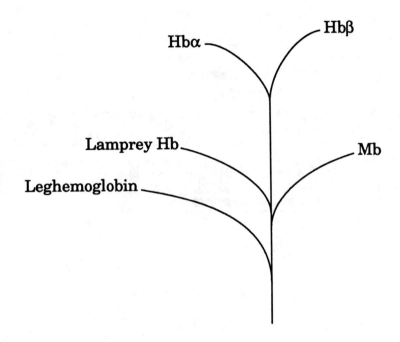

Since plants diverged from animals before animals diverged from each other, leghemoglobin must have diverged from the animal globins before they diverged from each other.

15. A 104-amino acid protein has 103 peptide bonds. Hence,

$$\% \text{ Yield} = (0.98 \times 0.97)^{103} \times 0.80 \times 100 = \textbf{0.43\%}$$

Chapter 7
THREE-DIMENSIONAL STRUCTURES OF PROTEINS

1. With 3.6 residues per turn and a pitch of 5.4 Å, the α-helix is 20 x 5.4/3.6 = **30Å** long.

   The end-to-end distance between two fully extended amino and residues is:

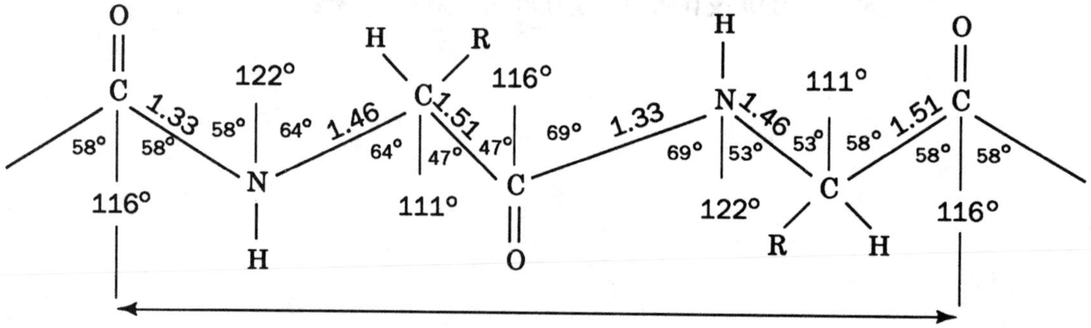

   1.33 sin 58°+ 1.46 sin 64°+ 1.51 sin 47°+ 1.33 sin 69°+ 1.46 sin 53 + 1.51 sin 58°= 7.23 Å

   Thus for 20 residues, this is 20 x 7.23 Å/2 = **72.3 Å.**

2. Near $\phi = 0°$, there is interference between the carbonyl oxygens of the two residues. Near $\phi = 60°$, the H atom substituents of the N and C_α atoms defining ϕ interfere. Past 90°, the carbonyl O interferes with the R group and then, towards 180°, with the H atom substituent of C_α. ϕ is relatively free of steric interference between about 240° (–120°) and 300° (–60°).

3. A 3_{10} helix incorporates two C_α atoms in a closed hydrogen bonded ring. Since a γ-amino acid residue also incorporates a β and γ carbon in its backbone, there must be 10 + 2 x 2 = 14 atoms in its hydrogen bonded ring analogous to a 3_{10} helix. With a pitch of 9.9 Å and a rise per residue of 3.2 Å, this helix must have 9.9/3.2 = 3.1 residues per turn. Hence, it is designated a 3.1_{14} helix.

4. (a) An α-helix
   (b) An antiparallel β pleated sheet, the two strands of which are connected by a reverse turn (residues 46-48) that resembles a type I β bend.
   (c) Glycine; no other type of residue can have these conformation angles.
   (d) Coil; these conformation angles have no obvious pattern.
   (e) Type I β bend. Note the Pro-Gly sequence.
   (f) The torsion angles about the rotatable bonds of the side chains are also required to define the three-dimensional structure of a protein.

5. Keratin should most easily split in a direction parallel to its fibrils as there are relatively few covalent bonds perpendicular to this direction that could hold the fibrils together. Thus, the fibrils must run parallel to hair fibers and across the fingers in fingernails.

6. α-Keratin is α helical so that its polypeptide chains can be stretched to their fully extended conformation before their covalent bonds are broken. Silk, which has a β-pleated sheet arrangement is nearly fully extended in its native conformation.

7.

$$\text{Growth rate} = \frac{15 \text{ cm/y} \times 10^8 \text{ Å/cm}}{365 \text{ d/y} \times 24 \text{ h/d} \times 3600 \text{ s/h}}$$

$$= 47.6 \text{ Å/s}$$

The coiled coil conformation of α-keratin reduces the normal 5.4 Å/turn pitch of α helices to 5.1 Å so that

$$\text{Growth rate} = \frac{47.6 \text{ Å/s}}{5.1 \text{ Å/turn}} = 9.3 \text{ turns/s}$$

8. No, because the bulky side chain of proline cannot fit in the extremely crowded central region of the collagen triple helix. Only a Gly side chain can fit there.

9. With 3.6 residues per turn, a 5-turn α-helix has 5 x 3.6 = 18 residues. Adjacent residues are 360°/3.6 = 100° apart on the helix and since a 19th residue would fall directly beneath the first, residues occur every 360°/18 = 20° in the helical wheel projection. The helical wheel is (in analogy with Figure 7-45):

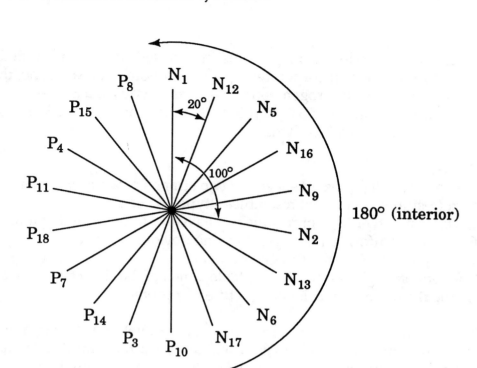

where N symbolizes nonpolar or internally hydrogen bonded uncharged polar residues and P represents charged or uncharged polar residues. Hence, the amino acid sequence is:

$$N_1\text{-}N_2\text{-}P_3\text{-}P_4\text{-}N_5\text{-}N_6\text{-}P_7\text{-}P_8\text{-}N_9\text{-}P_{10}\text{-}P_{11}\text{-}N_{12}\text{-}N_{13}\text{-}P_{14}\text{-}P_{15}\text{-}N_{16}\text{-}N_{17}\text{-}P_{18}$$

10. Lathyrogens are inhibitors of lysyl oxidase. This enzyme catalyzes the oxidation of lysine to allysine which participates in the covalent cross-linking of collagen. The administration of β-aminopropionitrile therefore impedes the formation of collagenous tissue such as scar tissue.

11. Carboxypeptidase A: An α/β protein with at least one βαβ unit.

   Triose phosphate isomerase: An α/β protein with multiple overlapping βαβ units.

   Myoglobin: An α protein with αα units.

   Concanavalin A: A β protein that has β meanders.

   Carbonic anhydrase: A β protein, or possibly an α + β protein, which has β meanders.

   Glyceraldehyde phosphate dehydrogenase:
   Lower domain (in Figure 7-46): An α/β protein with αβα units and two nucleotide binding folds.
   Upper domain: An α + β protein with a βαβ unit.

   Prealbumin: A β protein with several β meanders.

12. In one subunit there are 386 residues, each with a $(1–1/3000)$ probability of being correctly inserted. For the entire subunit this is

$$(1 – 1/3000)^{386} = 0.879$$

For the entire coat of 180 subunits, $180/0.879 = $ **205 subunits** will have to be synthesized, on the average, to form a perfect coat.

If the coat were one polypeptide chain, it would have $180 \times 386 = 69{,}480$ residues. The probability of such a coat being perfectly synthesized is $(1 – 1/3000)^{69480} = 8.71 \times 10^{-11}$. Consequently, $1/8.71 \times 10^{-11} = $ **1.14×10^{10}** coats would have to be synthesized in order to obtain an average of one perfect coat.

13. (a) C_5
    (b) D_4
    (c) D_2
    (d) D_3

14. Hemoglobin subunits have a high proportion of nonpolar residues in the contact regions between subunits. Analogous regions of myoglobin are exposed to solvent and are therefore largely polar. Consequently, hemoglobin has a higher ratio of nonpolar to polar residues than does myoglobin.

15. London dispersion forces are induced dipole-induced dipole interactions. A charge will induce an opposite charge on a nearby molecule through Coulombic interactions. Opposite charges attract.

16. Since they are normally immersed in a nonpolar environment, membrane-bound proteins will not be greatly affected by being coated by the nonpolar groups of detergents. Membrane-bound proteins are therefore resistant to detergent denaturation in comparison to normal globular proteins.

17. For proteins to aggregate, $\Delta G = \Delta H – T\Delta S$ must be negative. For hydrophobic bonding, ΔH is positive or zero and ΔS is positive. Although ΔG may be negative at some temperature, there may be a lower temperature at which the term $–T\Delta S$ has increased to the point that ΔG becomes positive. At this temperature, HbS disaggregates.

18. Both urea and guanidinium ion are likely to be strong hydrogen bonding agents. However, so is water, and it is usually in $55.5M$ concentration. Since water doesn't denature proteins, there is no reason to expect that urea or guanidinium ion can do so simply by competition for hydrogen bonds.

19. Causing a protein solution to foam greatly increases the area of the air-solution interface. Protein molecules at this interface have one side out of contract with water. Consequently, that side is not stabilized by hydrophobic bonding. Such protein molecules are destabilized so that they easily denature.

20. The following agents interfere with the hydrophobic effect: (a) Guanidinium chloride is a chaotropic agent. (b) Ethanol disrupts the hydrogen bonded structure of water. (d) Lower temperatures increase the entropic term for the free energy of spontaneously transferring a nonpolar group from water to a nonpolar medium ($\Delta G = \Delta H - T\Delta S$). (g) SDS incorporates nonpolar groups into micelles.

    The following agents do not necessarily interfere with the hydrophobic effect and therefore their dissociative effects do not support the hypothesis that the quaternary structure is stabilized only by hydrophobic bonding: (c) NaCl should not be disruptive of the hydrophobic effect since it is not chaotropic. However, its ions can disrupt salt bridges by acting as counterions for charged groups. (e) 2-Mercaptoethanol reduces disulfide bonds which might hold subunits together. (f) Changing the pH varies the charges of protein side groups thereby altering electrostatic interactions and possibly changing hydrogen bonds (perhaps this does not disprove that the quaternary structure is only stabilized by hydrophobic effects since changing the pH may only add a new disruptive influence).

21. There are equal amounts of monomers H and M. Upon dimerization, each H has an equal probability of combining with H or M so that it will form H_2 and HM in equal quantities. Similarly, M will dimerize to form equal quantities of HM and M_2. The ratio of these dimers is $H_2 : 2HM : M_2$. Each of these dimers has an equal probability of trimerizing with either H or M so that the distribution of trimers is $H_3 : 3H_2M : 3HM_2 : M_3$. Finally each of these trimers can hybridize with equal probability with H or M to form tetramers with the distribution $H_4 : 4H_3M : 6H_2M_2 : 4HM_3 : M_4$. [Note that each of these ratios are comprised of the terms in a binomial distribution, $(H + M)^n$ where $n = 1$, 2, 3 or 4].

22. Since SDS-polyacrylamide gel electrophoresis separates noncovalently linked subunits, the protein must consist of two types of subunits, A and B, with molecular masses of 10 and 17 kD, respectively. The largest cross-linked molecular mass of 74 kD indicates the protein has the subunit composition A_4B_2. The molecular masses of the cross-linked bands found correspond, in order of molecular masses, to A, B, A_2, AB, A_3, A_2B, A_4, A_3B, A_2B_2, A_4B, A_3B_2, A_4B_2. This indicates that all four A subunits are in contact (A_4), that the B subunits are not in contact (no B_2), and that the B subunits are interspersed by two A subunits. Thus, the geometry of the protein is

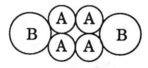

(athough the arrangement of the A's relative to one another is not apparent from the data provided.)

Chapter 8

PROTEIN FOLDING, DYNAMICS, AND STRUCTURAL EVOLUTION

1.

$$t = \frac{10^n}{10^{13}} \text{ s}$$

where n = number of peptide residues and the polypeptide backbone randomly reorients every 10^{-13} seconds.

For a 6-residue folding nucleus, $t = \dfrac{10^6}{10^{13}} = 10^{-7}$ s

For a 10-residue folding nucleus, $t = \dfrac{10^{10}}{10^{13}} = 10^{-3}$ s

For a 15-residue folding nucleus, $t = \dfrac{10^{15}}{10^{13}} = 100$ s

For a 20-residue folding nucleus, $t = \dfrac{10^{20}}{10^{13}} = 10^7$ s (about 4 months)

A polypeptide segment will be ineffective as a folding nucleus if it takes too long to fold. A 15-residue segment is probably a reasonable upper limit for the size of a folding nucleus.

2. (a) All of the 10 –SH groups in the protein form disulfides. There is only 1 way out of 9

that a particular –SH group can pair with its proper mate; only 1 way out of 7 that a particular one of the remaining 8 –SH's can pair with its proper mate, *etc.* Thus, the probability of all ten –SH groups randomly pairing with their respective proper mates is:

$$\frac{1}{9} \times \frac{1}{7} \times \frac{1}{5} \times \frac{1}{3} \times \frac{1}{1} = \frac{1}{945}$$

(b) Six of the 10 –SH groups in the protein form disulfides while four remain reduced. There are 10 ways to pick the first –SH group that doen't form a disulfide bond, 9 ways to pick the second, 8 ways to pick the third and 7 ways to pick the fourth. However, it is immaterial in which order they are picked: There are 4 ways an unbonded –SH group can be picked first, 3 ways that one can be picked second, and 2 ways that one can be picked third. Thus, the number of unique ways of picking four unbonded –SH groups out of a total of 10 –SH groups is

$$\frac{10 \times 9 \times 8 \times 7}{4 \times 3 \times 2} = 210$$

Of the remaining 6 –SH groups, there is only 1 out of 5 ways that a particular one of them will form an S—S bond with its proper mate; then only 1 way out of 3 possibilities that a particular one of the remaining –SH's will form the proper bond; *etc.* Hence, the probability of randomly forming 3 particular S—S bonds among 10 –SH groups is:

$$\frac{1}{210} \times \frac{1}{5 \times 3 \times 1} = \frac{1}{3150}$$

3. If a β-sheet were at the surface of a protein, its hydrogen bonds would be subject to competition from the aqueous solvent. This would destabilize the β-sheet. Solvent is largely excluded from the interior of the protein, which permits the β-sheet to form without this competition.

4. Lys has a charged side chain at physiological pH's. Thus, a polylysine α-helix would be bristling with mutually repelling positively charged groups which would destabilize the helix with respect to a more extended coil conformation in which Lys side chains are further apart. Increasing the pH until it is above the pK of the Lys residues (>10.8) neutralizes these charged groups thereby permitting polylysine to assume a more stable α-helix.

5.

							0.57	1.11	1.42	0.57	1.42	1.06	1.51
P_α													
α-Classification	*i*	*i*	*H*	*H*	*h*	*b*	*B*	*h*	*H*	*B*	*H*	*h*	*H*
Sequence	Arg$_{31}$-	Arg-	Glu-	Ala-	Gln-	Asn-	Pro-	Gln-	Ala-	Gly$_{40}$-	Ala-	Val-	Glu-
β-Classification	*i*	*i*	*B*	*i*	*h*	*i*	*B*	*h*	*i*	*b*	*i*	*H*	*B*
Hydropathy (Table 7-5)	–4.5	–4.5	–3.5	1.8	–3.5	–3.5	–1.6	–3.5	1.8	–0.4	1.8	4.2	–3.5

$\cdots \longrightarrow$ $\longleftarrow$ $\longrightarrow$

P_α	1.21	0.57			1.21	0.57	0.57	1.21	1.11	1.42	1.21	1.42	1.21	0.57	0.57
α-Classification	*H*	*B*	*B*	*B*	*H*	*B*	*B*	*H*	*h*	*H*	*H*	*H*	*H*	*H*	*B*
Sequence	-Leu	-Gly	-Gly	-Gly	-Leu	-Gly	-Gly$_{50}$	-Leu	-Gln	-Ala	-Leu	-Ala	-Leu	-Glu	-Gly-
β-Classification	*h*	*b*	*b*	*b*	*h*	*b*	*b*	*h*	*h*	*i*	*h*	*i*	*i*	*B*	*b*
Hydropathy	3.8	−0.4	−0.4	−0.4	3.8	−0.4	−0.4	3.8	−3.5	1.8	3.8	1.8	3.8	−3.8	−0.4

$\longleftarrow$ $\longrightarrow$

P_α					
α-Classification	*B*	*B*	*h*	*h*	*i*
Sequence	- Pro	- Pro$_{60}$	- Gln	- Lys	- Arg
β-Classification	*B*	*B*	*h*	*b*	*i*
Hydropathy	−1.6	−1.6	−3.5	−3.9	−4.5

$\longleftarrow$ $\longrightarrow$

According to the Chou-Fasman rules, residues 38 to 45 form an α-helix. Similarly, residues 50 to 57 form an α-helix. There is no segment of this sequence that can nucleate a β-sheet. The hydropathy is a minimum between residues 31 to 36 and there is no α-helix there. The same is true of residues 46 to 49 and 60 to 63. These regions are predicted to have reverse turns. Residues 38 to 57 therefore might form an αα-unit.

6. Taking the answer of Problem 7-9, the sequence of the proposed α-helix is:

$$N_1\text{-}N_2\text{-}P_3\text{-}P_4\text{-}N_5\text{-}N_6\text{-}P_7\text{-}P_8\text{-}N_9\text{-}P_{10}\text{-}P_{11}\text{-}N_{12}\text{-}N_{13}\text{-}P_{14}\text{-}P_{15}\text{-}N_{16}\text{-}N_{17}\text{-}P_{18}$$

where N represents a nonpolar residue and P represents a polar residue.

To ensure helix formation, let us make all of these residues helix formers (*H* or *h*). Referring to Table 8-1, N can be Ala, Ile, Leu, Met, Phe, Trp, or Val and P can be Gln, Glu, or Lys. Thus, one possibility is:

$$Ala_1\text{-}Ile\text{-}Gln\text{-}Glu\text{-}Leu_5\text{-}Met\text{-}Lys\text{-}Gln\text{-}Phe\text{-}Glu_{10}\text{-}Lys\text{-}Trp\text{-}Val\text{-}Gln\text{-}Glu_{15}\text{-}Ala\text{-}Ile\text{-}Lys_{18}$$

7. Folding nuclei are formed by the random folding of a polypeptide segment into a nativelike conformation. Since a β-sheet consists of several covalently distant polypeptide strands, such assemblies have insufficient frequencies of formation to be effective folding nuclei.

8. Folding nuclei, being of necessity small, are highly exposed to the surrounding aqueous solution. Any hydrogen bonds in such nuclei will be subject to competition with water molecules and therefore will not contribute stability to a folding polypeptide chain.

9. (a) Leu, being nonpolar, will most probably be located in the interior of a protein, a place that is efficiently packed. There would probably be no room for the bulkier Phe in place of the Leu so that such a substitution would be likely to disrupt the structure of the protein.

(b) Lys, being polar and positively charged, will almost certainly be a surface residue. Glu is also polar but is negatively charged and would therefore not contribute to any salt bridges in which Lys is involved. It would also change the overall charge on the protein. The packing of the protein would not be affected, however. Hence, the mutation would probably not greatly affect the protein's stability.

(c) A change of Val to Thr will place an H-bonding donor where none had previously been. It is therefore unlikely that a hydrogen bonding acceptor would be in the proper position to interact with the new Thr. Consequently, the native conformation of the mutant protein is likely to be less stable.

(d) Gly is often located at a position where the polypeptide chain bends in a way that sterically interferes with any residue side chain except that of Gly. Substitution of even Ala for Gly might prevent such a bend and thereby disrupt the protein structure.

(e) Pro is far more sterically constrained than is Met. Therefore, even though the two residues are reasonably similar in bulk and polarity, it is unlikely that the Pro could take up the same backbone conformation as the Met. Consequently, such a substitution would probably disrupt the protein structure.

10. Phe and Tyr are symmetrical for flips about their C_β—C_γ bonds; that is, the residue conformation is identical before and after the flip, including any hydrogen bond involving Tyr. Trp, however, is asymmetric for a ring flip, so that the protein conformation must differ before and after the flip and any hydrogen bond involving the Trp ring N—H must be broken by a ring flip. Equally important is that Trp is much more bulky than are Phe or Tyr so that a bigger cavity must form to permit Trp to flip than to permit Phe or Trp to flip. A large cavity is much less likely to form at random than a small cavity.

11. If the several known dehydrogenases independently evolved their coenzyme binding domains, this suggests that the structure of this domain is the only one that could successfully carry out the function of these enzymes. However, considering the great variety of known protein structures, it seems quite unlikely that this is true. Furthermore, even if the coenzyme binding domain were markedly superior to other types of structures in carrying out its function, it is improbable that the random processes of mutation would have independently "discovered" this complex fold on several occasions. Therefore, the hypothesis that the coenzyme binding domains of the various dehydrogenases arose by convergent evoluton must be rejected.

Chapter 9
HEMOGLOBIN: PROTEIN FUNCTION IN MICROCOSM

1. Since arterial blood in a normal individual at sea level is nearly fully saturated with O_2, hyperventilation can have little direct effect on the O_2 content of blood. However, the excessive removal of CO_2 would consume H^+ in the reaction

$$H^+ + HCO_3^- \rightleftharpoons H_2O + CO_2$$

The resultant higher blood pH would increase the O_2 affinity of Hb through the Bohr effect so that less than the normal amount of O_2 would be delivered to the tissues until the CO_2 balance was restored. Thus, hyperventilation might even have an effect opposite to what was intended. Furthermore, it is a dangerous procedure because repressing the breathing urge by lowering the CO_2 content of the blood may cause the diver to loose consciousness due to lack of O_2 and hence drown.

2. The derivation of the Hill equation assumes infinite cooperativity between all N ligand binding sites. Since there can be no larger cooperativity than this, n has its maximal value.

3. For the α N-termini of oxy Hb at pH 7.4,

$$RNH_3^+ \rightleftharpoons H^+ + RNH_2$$

$$K = 10^{-7.0} = \frac{[H^+][RNH_2]}{[RNH_3^+]}$$

$$\frac{[RNH_3^+]}{[RNH_2]} = \frac{[H^+]}{K} = \frac{10^{-7.4}}{10^{-7.0}} = 10^{-0.4} = 0.40$$

Hence, assuming a fraction x of the α-chains are in the RNH_3^+ state,

$$\frac{x}{1-x} = 0.40 \text{ so } x = 0.29.$$

Upon conversion to the deoxy state, each of Hb's α N-termini combine with

$$0.30 \times \frac{0.6 \text{ mol } H^+}{\text{mol } O_2} \times \frac{4 \text{ mol } O_2}{\text{mol Hb}} \times \frac{1 \text{ mol Hb}}{2 \text{ mol } \alpha \text{ N-termini}} = 0.36 \frac{\text{mol } H^+}{\text{mol } \alpha \text{ N-termini}}$$

so that the fraction of the α-chains in the RNH_3^+ state is now $0.29 + 0.36 = 0.65$. Hence,

$$pK = -\log\left\{\frac{[H^+][RNH_2]}{[RNH_3^+]}\right\}$$

$$= -\log\left\{10^{-7.4}\left(\frac{1-0.65}{0.65}\right)\right\} = 7.67$$

4. According to Figure 9-9, a 48 hour stay at sea level after adaptation lowers the p_{50} for O_2 in blood beyond the ability of a 24 hour readaptation to restore this p_{50} to its original adaptation level. Hence, this could be an attempt to handicap you.

5. For Mb, according to equation [9.4]

$$Y_{O_2} = \frac{pO_2}{p_{50} + pO_2} \qquad \text{where } p_{50} = 2.8 \text{ torr}$$

At 10 torr, $Y_{O_2} = \dfrac{10}{2.8 + 10} = 0.78$

At 1 torr, $Y_{O_2} = \dfrac{1}{2.8 + 1} = 0.26$

so that $\Delta Y_{O_2} = 0.78 - 0.26 = 0.52$

Therefore, across the pO_2 gradient in active muscle, Mb can transport a significant amount of O_2 by diffusion (far more so than could be transported in the absence of Mb due to the insolubility of O_2 in H_2O). However, in the absence of a steep pO_2 gradient spanning its p_{50}, ΔY_{O2} for Mb is small due to the hyperbolic shape of its binding curve so that Mb would be ineffective as an O_2 transport protein in slowly metabolizing tissues.

6. Depletion of BPG shifts Hb's fractional saturation curve to the left (lowers p_{50}) so that the amount of O_2 that Hb can deliver to the tissues (ΔY_{O_2}) is reduced. Thus, week old blood is not as effective as fresh blood for use in transfusions.

7. Equation [9.12], the Hill equation for O_2, states

$$\log\left(\frac{Y_{O_2}}{1-Y_{O_2}}\right) = n \log pO_2 - n \log p_{50}$$

so that a plot of $\log\left(\dfrac{Y_{O_2}}{1-Y_{O_2}}\right)$ *vs* $\log pO_2$ should have a slope of n and an intercept on

the $\log pO_2$ axis of p_{50}.

pO_2	$\log pO_2$	$\dfrac{Y_{O_2}}{1-Y_{O_2}}$	$\log\left(\dfrac{Y_{O_2}}{1-Y_{O_2}}\right)$
20	1.30	0.16	−0.80
30	1.48	0.35	−0.46
40	1.60	0.64	−0.19
50	1.70	1.00	0
60	1.78	1.44	0.16
70	1.84	1.94	0.29
80	1.90	2.57	0.41
90	1.95	3.17	0.50

This is plotted below:

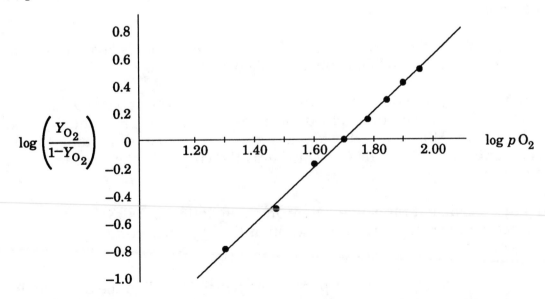

This linear plot has a slope, the Hill constant, of $n = 2$ and an intercept on the $\log pO_2$ axis of $\log p_{50} = 1.70$ so that

$$p_{50} = 10^{1.70} = \textbf{50 torr}$$

Normal Hb has a Hill constant of ~3 and a p_{50} of 26 torr in fresh blood, so the blood sample analyzed here is **abnormal.**

8. In the case of anemia, the Hb that is present functions normally and is presumably present in sufficient quantity to carry the required amount of O_2 (at least under conditions of low exertion). In the CO poisoning case, half the O_2 binding sites of Hb bind CO essentially irreversibly. This converts most of the Hb to the R state which greatly increases its O_2 affinity over that of normal Hb. Therefore, at the tissues, little of the O_2 carried by this Hb will be discharged resulting in the asphyxiation of the victim.

9. With the β C-termini held in abnormal position, they will be unable to form their normal salt bridges in the T form of Hb. These salt bridges stabilize the T form of Hb so that the R form of Hb Rainier is favored more than the R form of HbA.

   (a)   The O_2 affinity of the R state is greater than that of the T state. Hence, the O_2 affinity of Hb Rainier is greater than that of HbA (p_{50} increases).

   (b)   The Bohr effect will be approximately halved since the salt bridges involving His 146β, which absorb a proton upon the deoxygenation of HbA, cannot form in Hb Rainier (the Bohr effect in Hb Rainier is experimentally determined to be half that of HbA).

   (c)   The O_2 binding cooperativity is much reduced in Hb Rainier because its R state is more stable in the deoxygenated molecule than it is in HbA. Accordingly, the Hill constant is greatly reduced from its normal value of abut 3.0 (the Hill constant of Hb Rainier is experimentally found to be 1.5).

   (d)   BPG binds to the T state of Hb. Since the T state in Hb Rainier is less stable than that in HbA, the BPG affinity of Hb Rainier must be likewise reduced.

10. As the crocodile remains underwater without breathing, its metabolism generates CO_2 and hence the HCO_3^- content of its blood increases. The HCO_3^- preferentially binds to the crocodile's deoxyHb, which allosterically prompts the Hb to assume the deoxy conformation and thus release its O_2. This helps the crocodile stay underwater long enough to drown its prey.

11. The HbA must participate to some extent in the gelation of HbS. If not, it would probably increase rather than reduce the gelation time by interfering with interactions between HbS molecules.

12. The elevated BPG in sickle cell anemia permits the remaining Hb in the blood to deliver more O_2 to the tissues than if the BPG level were normal. However, since BPG stabilizes the T form of Hb, it promotes sickling and therefore aggrevates the disease. Hence, elevated BPG is a mixed blessing in sickle-cell anemia.

13. These mutant Hb's have the following properties:

    (1) Heterozygotes for HbS at sea level are clinically normal. However, it is possible that their erythrocytes will sickle at very high altitudes under conditions of great exertion.

    (2) Hb Hyde Park (which is the analogous mutation on the β chain to Hb Iwate; Section 9-3A) is an HbM since the mutant Tyr in place of the proximal His ligands the Fe and stabilizes it in the Fe(III) (met) form. These met subunits cannot bind O_2. Hence, this heterozygous blood can only bind about 3/4 of the O_2 that normal blood can. Its cooperativity must also be reduced.

    (3) Hb Riverdale-Bronx is similar to Hb Savannah (Section 9-3A). It places a bulky group in the tertiary structure of the β chain where only a Gly will fit. Hence, these subunits are unstable and the mutant climber suffers from hemolytic anemia.

    (4) Hb Memphis is a surface mutation (see Figure 9-13) that substitutes one polar residue for another of similar bulk. Hence, this mutation is harmless.

    (5) Hb Cowtown eliminates a C-terminal salt bridge holding deoxy Hb in the T state (see Fig. 9-18b). Hence, this mutation decreases the stability of the T state thereby increasing Hb's O_2 affinity and decreasing its cooperativity.

    Of all the four candidates, the one with Hb Memphis is most likely not to have serious physiological difficulties at high altitudes and would therefore be the best choice for the expedition.

14. In the Adair equation (equation [9.17]) for $k_1 \approx k_2 \approx k_3 >> k_4$, the only variable terms that are significant contain k_4.

    Hence,

    $$Y_S = \frac{[S]^4/k_1k_2k_3k_4}{1 + [S]^4/k_1k_2k_3k_4} = \frac{[S]^4}{k_1k_2k_3k_4 + [S]^4}$$

    This is the Hill equation with $K = k_1k_2k_3k_4$.

    If $k_1 = k_2 = k_3 = k_4$, the numerator and denominator of equation [9.17] can be factored to yield:

    $$Y_S = \frac{\dfrac{[S]}{k_1}\left(1 + \dfrac{[S]}{k_1}\right)^3}{\left(1 + \dfrac{[S]}{k_1}\right)^4} = \frac{[S]}{k_1 + [S]}$$

which describes a hyperbola.

15.
$$R_0 \rightleftharpoons T_0$$

$$R_0 + S \rightleftharpoons R_1 \qquad\qquad T_0 + S \rightleftharpoons T_1$$
$$R_1 + S \rightleftharpoons R_2 \qquad\qquad T_1 + S \rightleftharpoons T_2$$

$$L = [T_0]/[R_0] \qquad c = k_R/k_T \qquad \alpha = [S]/k_R$$

$$[R_1] = [R_0]\left(\frac{n-1+1}{1}\right)[S]/k_R \qquad\qquad [T_1] = [T_0]\left(\frac{n-1+1}{1}\right)[S]/k_T$$

$$[R_2] = [R_1]\left(\frac{n-2+1}{2}\right)[S]/k_R \qquad\qquad [T_2] = [T_1]\left(\frac{n-2+1}{2}\right)[S]/k_T$$

$$= [R_0]\frac{n(n-1)}{2}[S]^2/k_R^2 \qquad\qquad = [T_0]\frac{n(n-1)}{2}[S]^2/k_T^2$$

$$K_2 = \frac{[T_2]}{[R_2]} = \frac{[T_0]}{[R_0]}\left(\frac{k_R}{k_T}\right)^2$$

$$K_2 = Lc^2$$

16.
$$\overline{R} = \frac{[R_0] + [R_1] + \dots + [R_n]}{([R_0] + [R_1] + \dots + [R_n]) + ([T_0] + [T_1] + \dots + [T_n])}$$

But substituting $[R_{n-1}]$ for $[R_n]$, $[R_{n-2}]$ for $[R_{n-1}]$, *etc.,*

$$[R_0] + [R_1] + \dots + [R_n] = [R_0]\left(n\alpha + \frac{n(n-1)\alpha}{2} + \dots + \frac{n!\,\alpha^n}{n!}\right)$$

$$= [R_0](1 + \alpha)$$

Similarly

$$[T_0] + [T_1] + \dots + [T_n] = [T_0](1 + [S]/k_T)^n = L[R_0](1 + c\alpha)^n$$

so that

$$\overline{R} = \frac{(1+\alpha)^n}{(1+\alpha)^n + L(1+c\alpha)^n}$$

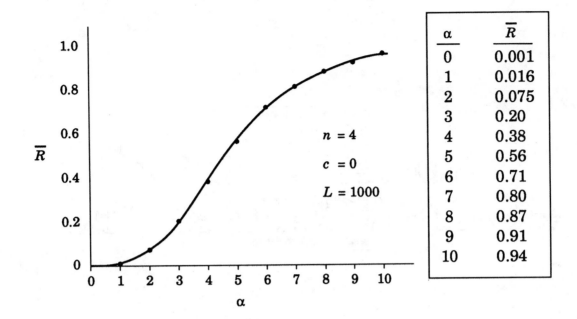

At $c = 0$, S binds only to the R state; at $L = 1000$, the R state is highly unfavorable. The $\overline{R}$ *vs* α curve is sigmoidal indicating cooperativity. Only above $\alpha = 2$ is there significant binding of S to R. However, once some S is bound to R this tends to hold the protein in the R state so it is much easier for additional S to bind to it. This stabilizes the R state to an even greater degree thereby resulting in the cooperativity of binding of S to R.

17. Inhibitor binds only to the T state, to which substrate cannot bind. By binding to the protein and thus forcing it to assume the T state, the inhibitor makes it easier for more inhibitor to bind to the protein. Hence, this is a homotropic effect in exactly the same manner that binding S to the R state when S doesn't bind to the T state is a homotropic effect.

Chapter 10
SUGARS AND POLYSACCHARIDES

1. O-α-D-galactopyranosyl-(1 → 6)-O-α-D-glucopyranosyl-(1 → 2)-β-D-fructofuranoside. It has no free anomeric hydroxyl group so that it is not a reducing sugar.

2.

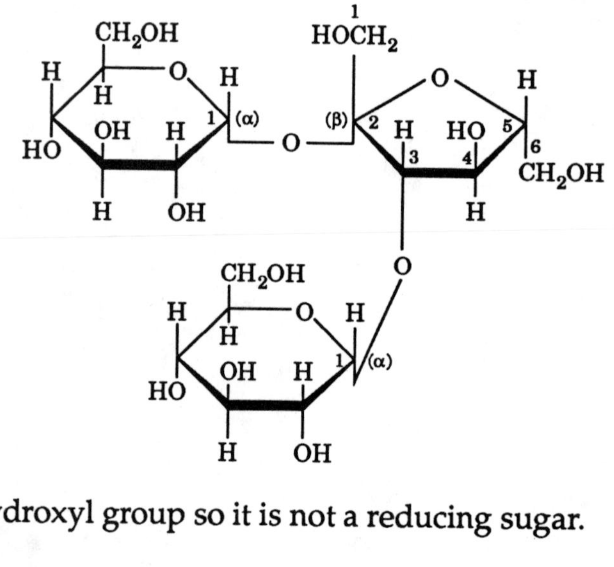

It has no anomeric hydroxyl group so it is not a reducing sugar.

3. (2R,3S,4R,5R) -2,3,4,5,6-pentahydroxy-1-hexanal.

4.

α-D-Talofuranose

β-L-Sorbopyranose

5.

```
   CHO                 CH2OH                          CH2OH              CH2OH
   |                   |                              |                  |
H—C—OH             H—C—OH                         H—C—OH             C=O
   |                   |                              |                  |
HO—C—H    NaBH4    HO—C—H     Stereospecific     HO—C—H            HO—C—H
   |       ——→         |       reduction            |           ≡      |
H—C—OH             H—C—OH        ←——            H—C—OH             H—C—OH
   |                   |                              |                  |
H—C—OH             H—C—OH                          C=O             HO—C—H
   |                   |                              |                  |
   CH2OH               CH2OH                         CH2OH              CH2OH

D-Glucose        D-glucitol or L-sorbitol                     L-Sorbose
```

Reduction of D-glucose or stereospecific reduction of L-sorbose, as shown, yields the same product which is therefore D-glucitol or L-sorbital.

6. All saccharide linkages involve a glycoside bond, that is, C(1). For a disaccharide, C(1) of the glycosidic glucose may be α or β. It may link to C(1), C(2), C(3), C(4) or C(6) of the other glucose while the latter's C(1) is α or β. However, the C(1)α—C(1)β link gives the identical compound as the C(1)β—C(1)α link. Hence there are 4 x 5 – 1 = **19** possible glucose disaccharides.

For each glucose disaccharide, there are 8 positions to which a third glucose can form a glycosidic bond. This glycosidic bond may also be α or β. Hence, there are 2 x 8 x 19 = **304** possible glucose trisaccharides.

7. Each molecule of amylopectin has only **one reducing end.**

8. In the absence of lignins, the cellular fibers are largely held together by hydrogen bonds between their glucose residues. Moreover, the absence of lignins makes these hydrogen bonds far more accessible than they are in wood. Thus water, a powerful hydrogen bonding donor and acceptor, competes for paper's interfiber hydrogen bonds and thereby weakens them. Oil, which has few, if any, hydrogen bonding groups, has no such effect.

9.

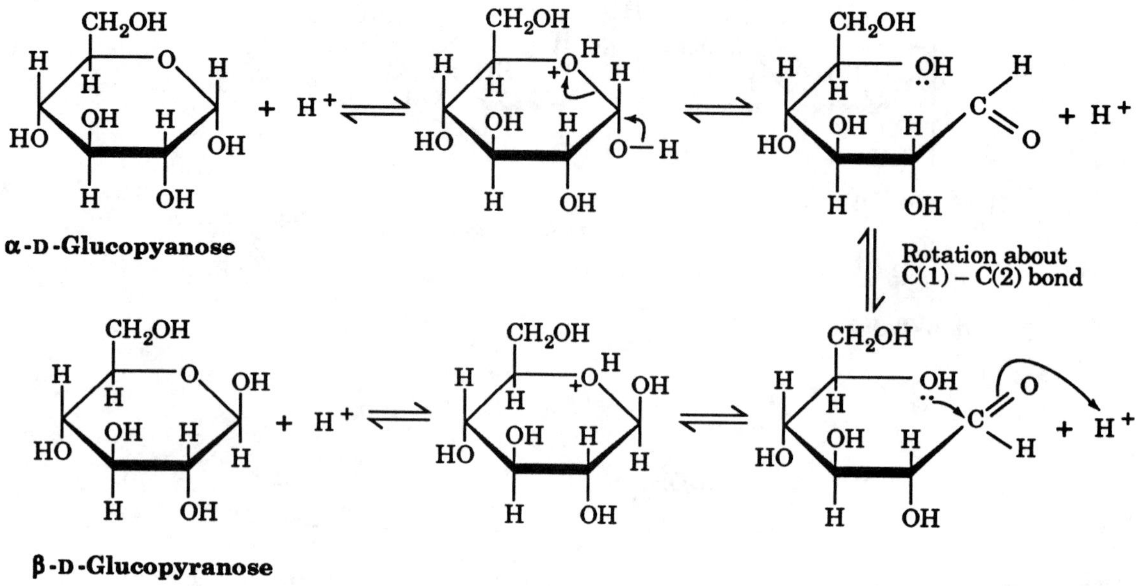

α-D-Glucopyanose
Rotation about
C(1) – C(2) bond
β-D-Glucopyranose

10. For the initial mixture:

$$[\alpha]_D^{20} = (0.20 \times 150.7) + (0.80 \times 52.8) = \mathbf{72.4^\circ}$$

For the equilibrium mixture, if x is the fraction of the α-anomer, then $1 - x$ is the fraction of the β-anomer and

$$150.7x + 52.8\,(1 - x) = 80.2$$

$$x = 0.28;\ 1 - x = 0.72$$

The equilibrium mixture is **28% α-anomer and 72% β-anomer.**

11. A monosaccharide's epimers are related to it by inversion about any non-anomeric asymmetric center. Hence, for D-gulose they are D-idose, D-galactose and D-allose. (see Figure 10-1).

12.

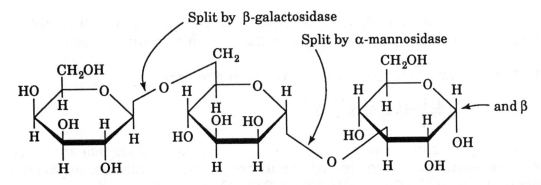

***O*-β-D-galactopyranosyl-(1→6)-*O*-α-D-mannopyranosyl-(1→3)-D-glucopyranose**

13.

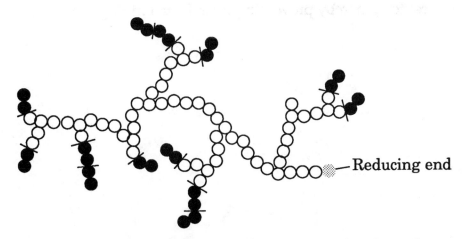

The filled circles represent the glucose residues that can be cleaved away by β-amylase. The remainder is the limit dextrin.

14. The ingestion of an effective intestinal α-amylase inhibitor with a starch-containing meal would result in the starch being transported, in undigested form, to the colon. There, the bacterial flora, which are equipped with a large variety of carbohydrate-hydrolyzing enzymes, would readily digest it, with a resulting digestive upset similar to but probably more severe than that suffered by individuals with milk intolerance who drink milk. However, since the purported α-amylase inhibitor is a protein, it would probably be denatured and partially degraded by the high acidity and (acid-resistant) proteolytic enzymes in the stomach before it enters the intestines.

15. A glucosyl residue has a molecular mass of 162 D. Hence, 6.0 g of glycogen represents 37.0 mmol of glucose residues.

   (a) The fraction of glucose residues at branch points is

$$\frac{3.1}{37.0} = 0.0837$$

Hence, the average number of glucose residues per branch is 1/0.0837 = **11.9.**

(b) 2,3,6-Tri-*O*-methylglucose (the product of the glucose residues not at the branch points). There should be 37.0 – 3.1 = 33.9 mmol of this product present.

(c) The fraction of glucose residues at reducing ends is:

$$\frac{0.0031}{37} = 8.38 \times 10^{-5}$$

This corresponds to one out of 1/8.38 x 10^{-5} = 11,900 residues that are at reducing ends. Since there can be only one reducing end per glycogen molecule, the average molecular mass of the sample is 162 D x 11,900 = **1,930,000 D.**

16. *E. coli* bind *O*-mannose residues on cell surfaces. Methyl-α-D-mannoside will also bind to these *E. coli* receptors, thereby preventing them from binding to the cell surface residues.

Chapter 11
LIPIDS AND MEMBRANES

1. A cis double bond puts a permanent 120° bend in a hydrocarbon chain whereas, with a trans double bond, such a bend closely resembles the shape of a fully extended saturated chain. Hence, trans double bonds do not interfere with the packing of hydrocarbon chains to form a crystal as much as cis double bonds so that the former have the higher melting point.

2. Animals in cold climates generally insulate themselves with a layer of fat (blubber). This usually has a high proportion of polyunsaturated fatty acids which prevents the fat from solidifying.

3. Each of the 2 fatty acid attachments in phosphatidylserine can have one of the four different fatty acid residues for a total of 4 x 4 = **16 different phosphatidylserines.** In triacylglycerols, only 4 x 5/2 = 10 of the 16 pairs of substituents at C(1) and C(3) are unique because molecules with different substituents at C(1) and C(3) are identical to the molecule with the reverse substitution order. In addition, the C(2) position may have any of the four substituents for a total of 4 x 10 = **40 different triacylglycerols.** Cardiolipins consist of two phosphatidic acid residues linked together by a glycerol moiety. Hence, cardiolipins with the same pair of fatty acids on both phosphoglycerides are identical to those with this order reversed. Thus, there are 16 x 17/2 = **136 different cardiolipins.**

4. According to Franklin, about 1 teaspoon of oil spread over about half an acre. Since

$$x = \frac{V}{A}$$

where x = thickness of the oil layer, V = volume of oil, and A = area of the oil. Then

$$x = \frac{5 \, cm^3 \times (10^8 \, \text{Å} / cm)^3}{4047 \, m^2 / 2 \times (10^{10} \, \text{Å} / m)^2} = 25 \, \text{Å}$$

5. Ca^{2+} salts of soaps are insoluble. Consequently, adding soap to hard water simply results in Ca^{2+} precipitates of the soap which is therefore not available to act as an amphiphile.

6. Hydrocarbons are entirely hydrophobic, not amphiphilic. Hence, they tend to minimize their contact with water, not spread over it in a monolayer as do amphiphiles.

7. Air is nonpolar. Consequently, the hydrophobic tails of soap ions extend into the air to minimize their contact with the polar aqueous interior of the soap film. The anionic head groups of the soap remain solvated in the water.

8. Detergents form micelles into which proteins, which are amphiphilic molecules, can be dispersed. The hydrocarbon portions of the detergent molecules provide a hydrophobic environment in the interior of the micelle that prevents integral membrane proteins from hydrophobically aggregating and precipitating. Mild detergents, such as Triton X-100, are not sufficiently amphiphilic to disrupt the hydrophobic forces holding a water soluble protein in its native conformation. Since these proteins have polar exteriors they do not readily interact with such detergents. However, integral membrane proteins are coated by such detergents over the hydrophobic portions of their exteriors.

9. The following table contains the trans-membrane sequence of glycophorin A together with its leading and trailing tripeptide. The helix-forming propensities are given according to the Chou and Fasman rules.

Residue:	-Glu[70]- Pro - Glu - Ile - Thr - Leu - Ile - Ile - Phe - Gly - Val[80]- Met-Ala-
Propensity:	H B H h i H h h h B h H H
P_α:	1.51 0.57 1.51 1.08 0.83 1.21 1.08 1.08 1.13 0.57 1.06 1.45 1.42

Residue:	Gly - Val - Ile - Gly - Thr - Ile - Leu - Leu[90]- Ile - Ser - Tyr - Gly
Propensity:	B h h B i h H H h i b B
P_α:	0.57 1.06 1.08 0.57 0.83 1.08 1.21 1.21 1.08 0.77 0.69 0.57

The underlined portions of the sequence satisfy the Chou and Fasman criteria for α-helix formation and the remaining residues are not far below them in helix forming propensity. Evidently, the trans-membrane sequence is mostly, if not entirely, α-helical.

10. Since all subunits must have the same orientation with respect to the membrane, the only allowed rotational symmetry axes must be perpendicular to the membrane plane. This constraint eliminates all but cyclic symmetries; that is, C_2, C_3, C_4, etc.

11. The A-antigen differs from the H-antigen by the addition of an *N*-acetylgalactosamine terminal residue. This residue is galactose in the B-antigen. *N*-Acetylgalactosamine binds to the anti-A antibody in its binding site for the *N*-acetylgalactosamine residue, thereby blocking the binding of this residue to the A-antigen. The same is true of galactose for the anti-B antibody.

12. Type O individuals are universal donors because their red cells do not carry A or B antigens. Hence, the anti-A antibodies in the plasma of types B and O individuals or the anti-B antibodies in the plasma of types A and O individuals will not agglutinate these cells. Type AB individuals are the universal recipients because their blood plasma contains neither anti-A nor anti-B antibodies so that it will not agglutinate the red cells from donors with other blood types.

13. Anti-H antibodies would be expected to similarly agglutinate cells from the tissues of individuals with A, B and O blood types. This is because they all contain the H-antigen (which in the A and B-antigens, respectively, have appended *N*-acetylgalactosamine and galactose residues).

14. A model that accounts for these observations is that the LDL-receptor glycoprotein is synthesized and randomly inserted into the plasma membrane as are other membrane proteins. The receptor protein then laterally diffuses to a coated pit where it binds to clathrin (or to a protein that is bound to clathrin) via a site that is distinct from the LDL binding site. This process is apparently defective in the form of familial hypercholesterolemia under consideration.

15. Lipoproteins are oil droplets (triacylglycerols and cholesteryl esters) enclosed by a monomolecular layer of phospholipids, cholesterol and proteins. Since the surface-to-volume ratio of a particle decreases with its diameter, the proportion of the relatively dense outer layer molecules in the lipoprotein particle increases as its diameter decreases.

16.

In this model of virus formation the protein shell is inside the membrane of the mature virus particle.

17. Water insoluble cholesteryl esters are formed in cells that have an excess of cholesterol. In normal cells, these cholesteryl esters are eventually mobilized by hydrolysis, as catalyzed by cholesteryl ester hydrolase, and the resulting cholesterol is incorporated into the cell's membranes. In the cells of individuals with Wolman's disease, the cholesteryl esters collect as numerous inclusions (Wolman's disease is invariably fatal in infancy).

Chapter 12
INTRODUCTION TO ENZYMES

1.

$$\underset{\substack{\| \\ O}}{CH_3CH} + NADH + D^+ \longrightarrow CH_3CH_2OD + NAD^+$$

where the D^+ is supplied by the D_2O solvent.

2. YADH catalyzes the oxidation of methanol to formaldehyde:

$$CH_3OH + NAD^+ \longrightarrow \underset{\substack{\| \\ O}}{HCH} + NADH + H^+$$

with the hydrogen from the methyl group being transferred to the NADH.

YADH is unable to differentiate the various hydrogen isotopes in (R)-TDHCOH. Consequently, the reaction will occur with the following equimolar stoichiometries:

$$NAD^+ + (R)\text{-TDHCOH} \Bigg\langle \begin{array}{l} \longrightarrow \; T-\overset{\displaystyle O}{\overset{\|}{C}}-D \; + \; NADH \; + \; H^+ \\[2ex] \longrightarrow \; T-\overset{\displaystyle O}{\overset{\|}{C}}-H \; + \; NADD \; + \; H^+ \\[2ex] \longrightarrow \; D-\overset{\displaystyle O}{\overset{\|}{C}}-H \; + \; NADT \; + \; H^+ \end{array}$$

Here the NADD and the NADT have the chiralities indicated in Section 12-2A.

3. Maleate, being geometrically distinct from fumarate, would not be expected to bind to fumarase. Hence maleate most probably does not react in the presence of fumarase.

4.

$$R-\overset{\displaystyle O}{\overset{\|}{C}}-OR' \; + \; H_3\overset{+}{N}-\overset{\displaystyle R''}{\overset{|}{C}H}-COO^-$$

$$\downarrow$$

$$R-\overset{\displaystyle O}{\overset{\|}{C}}-NH-\overset{\displaystyle R''}{\overset{|}{C}H}-COO^- \; + \; HOR' \; + \; H^+$$

5. The NaOH treatment of corn makes its nicotinamide nutrionally available. This prevents pellagra on a corn rich diet such as that once common in rural Southern United States.

6. The CTP exhibits the greatest cooperativity (sigmoidicity). At low [aspartate], CTP-ATCase is least reactive but at high [aspartate] has become the most active. Thus, aspartate binding activates CTP-ATCase more than ATP-ATCase or ATCase alone.

7. This would enable the control of the pathway. An excess of the final product would slow the rate of synthesis of this product through feedback inhibition. Moreover, since it is the first step of the pathway that is shut down, there would be no unnecessary (and perhaps deleterious) accumulation of pathway intermediates.

8. Catalase. Hydrogen-peroxide:hydrogen-peroxide oxidoreductase, E.C. 1.11.1.6.

Aspartate carbamoyltransferase. Carbamoylphosphate:L-aspartate carbamoyltransferase, E.C. 2.1.3.2.

Trypsin has no systematic name. E.C. 3.4.21.4.

Chapter 13
RATES OF ENZYMATIC REACTIONS

1.

Time (min)	[Sucrose] (M)	ln [Sucrose]
0	0.5011	−0.691
30	0.4511	−0.796
60	0.4038	−0.907
90	0.3626	−1.014
130	0.3148	−1.156
180	0.2674	−1.319

Plotting this data:

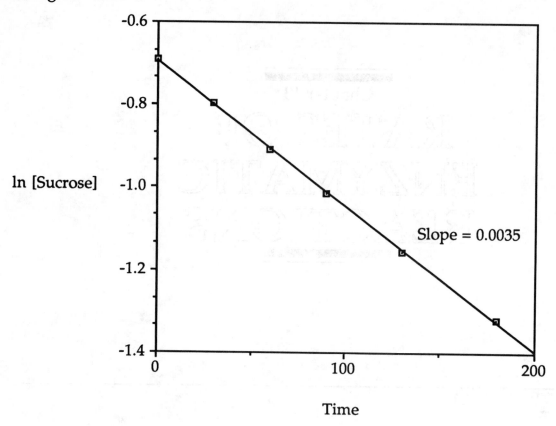

$$k = -\text{slope} = 0.0035 \text{ min}^{-1}$$

$$t_{1/2} = \frac{0.693}{k} = \frac{0.693}{0.0035 \text{ min}^{-1}} = 198 \text{ min}$$

The reaction follows pseudo first-order kinetics because in aqueous solution the H_2O concentration of $55.5M$ remains essentially constant in comparison to the sucrose concentration. The reaction actually is second-order.

$$(1 - 0.99) = 0.01 = (1/2)^n \qquad \text{where } n \text{ is the number of half-lives.}$$

$$n \log 1/2 = \log 0.01$$

$$n = \frac{-2.00}{-0.301} = 6.64 \text{ half-lives}$$

Time to hydrolyze 99% of sucrose = $t_{0.01} = n t_{1/2} = 6.64 \times 198 = $ **1315 min**

The time to react a given fraction of a substance that reacts in a first-order process is independent of the amount initially present. Hence, the time to hydrolyze 99% of 2 x 0.5011M sucrose is also **1315 min.**

2.

$$\frac{d[P]}{dt} = k'e^{-\Delta G^{\ddagger}/RT}[A][B]$$

If $\Delta G^{\ddagger}_{old} - \Delta G^{\ddagger}_{new} = \Delta\Delta G^{\ddagger}$

$$\frac{d[P]_{new}}{dt} = k'e^{-\Delta G^{\ddagger}_{new}/RT}[A][B]$$

$$= k'e^{-(\Delta G^{\ddagger}_{old} - \Delta\Delta G^{\ddagger})/RT}[A][B]$$

$$= \frac{d[P]_{old}}{dt} e^{\Delta\Delta G^{\ddagger}/RT}$$

For $T = 25°C = 273 + 25 = 298$ K and $\Delta\Delta G^{\ddagger} = 1$ kJ•mol^{-1} = 1000 J•mol^{-1},

$$\frac{d[P]_{new}}{dt} = e^{(1000\,J•mol^{-1}/8.314\,J•mol^{-1}K^{-1} \times 298\,K)}\frac{d[P]_{old}}{dt} = 1.5\,\frac{d[P]_{old}}{dt}$$

For $\Delta\Delta G^{\ddagger} = 10$ kJ•mol^{-1} = 10,000 J•mol^{-1},

$$\frac{d[P]_{new}}{dt} = e^{(10,000\,J•mol^{-1}/8.314\,J•mol^{-1}K^{-1} \times 298\,K)}\frac{d[P]_{old}}{dt} = 57\,\frac{d[P]_{old}}{dt}$$

3.

$$K_S = \frac{k_{-1}}{k_1} = \frac{2 \times 10^4\,s^{-1}}{5 \times 10^7\,M^{-1}•s^{-1}} = 4 \times 10^{-4}\,M$$

$$K_M = \frac{k_{-1} + k_2}{k_1} = \frac{(2 \times 10^4 + 4 \times 10^2)\,s^{-1}}{5 \times 10^7\,M^{-1}•s^{-1}} = \frac{2.04 \times 10^4\,s^{-1}}{5 \times 10^7\,M^{-1}•s^{-1}} \approx 4 \times 10^{-4}\,M$$

Since $K_M \approx K_S$ here, the substrate binding will achieve equilibrium as well as steady state in this case.

4. (a)

[S] (mM)	1/[S] (M^{-1} x 10^{-3})	(1)v_O (μM•s^{-1})	(1) 1/v_O (M^{-1}•s x 10^{-6})
1	1	2.5	0.40
2	0.5	4.0	0.25
5	0.2	6.3	0.16
10	0.1	7.6	0.13
20	0.05	9.0	0.11

(2) $v_O(\mu M \bullet s^{-1})$	(2) $1/v_O(M^{-1}\bullet s \times 10^{-6})$	(3) $v_O(\mu M \bullet s^{-1})$	(3) $1/v_O (M^{-1}\bullet s \times 10^{-6})$
1.17	0.85	0.77	1.30
2.10	0.48	1.25	0.80
4.00	0.25	2.00	0.50
5.7	0.18	2.50	0.40
7.2	0.14	2.86	0.35

Plotting $1/v_o$ vs $1/[S]$ for these data:

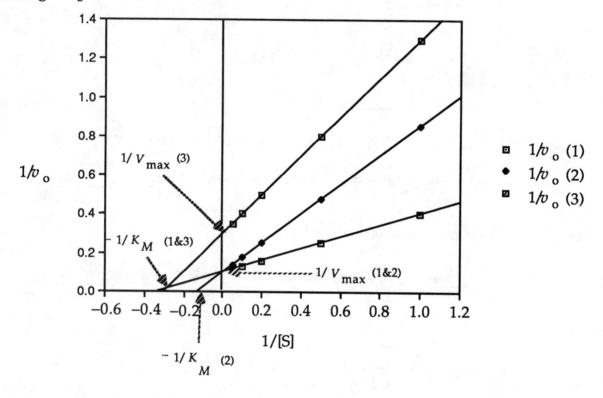

(1) $\dfrac{-1}{K_M} = -0.3 \times 10^{-3} M^{-1}$ $\dfrac{1}{V_{max}} = 0.1 \times 10^{6} M^{-1}\bullet s$

$K_M = 3.3 \times 10^{-3} M = \mathbf{3.3\ mM}$ $V_{max} = 10^{-5} M\bullet s^{-1} = \mathbf{10\ \mu M \bullet s^{-1}}$

(2) $\dfrac{-1}{K_M^{app}(2)} = -0.13 \times 10^{3} M^{-1}$ $V_{max}^{app}(2) = \mathbf{10\ \mu M \bullet s^{-1}}$ [same as (1)]

$K_M^{app}(2) = 7.7 \times 10^{-3} M = \mathbf{7.7\ mM}$

(3) $K_M^{app}(3) = \mathbf{3.3\ mM}$ [same as (1)] $V_{max}^{app}(3) = 3.3 \times 10^{-6} M\bullet s^{-1} = \mathbf{3.3\ \mu M \bullet s^{-1}}$

Curve (2) intersects curve (1) on the $1/v_O$ axis. Hence, inhibitor (1) acts **competitively.**

$$K_M^{app}(2) = \alpha K_M$$

$$7.7 \text{ m}M = \alpha 3.3 \text{ m}M$$

$$\alpha = \frac{7.7 \text{ m}M}{3.3 \text{ m}M} = 2.33 = \left(1 + \frac{[I]}{K_I}\right)$$

where $[I] = 10 \text{ m}M = 0.01M$.

$$K_I = \frac{[I]}{\alpha - 1} = \frac{0.01M}{2.33 - 1} = 0.0075M = 7.5 \text{ m}M$$

Curve (3) intersects curve (1) on the $-1/[S]$ axis. This is diagnostic of **mixed inhibition** (with $\alpha = \alpha'$).

$$\frac{1}{V_{max}^{app}(3)} = \frac{\alpha'}{V_{max}}$$

$$\alpha' = \frac{V_{max}}{V_{max}^{app}(3)} = \frac{10 \text{ }\mu M \bullet s^{-1}}{3.3 \text{ }\mu M \bullet s^{-1}} = 3.00 = \left(1 + \frac{[I]}{K'_i}\right)$$

$$[I] = 10 \text{ m}M = 0.01M$$

$$K'_I = K_I = \frac{[I]}{\alpha' - 1} = \frac{0.01M}{3.00 - 1} = 5 \times 10^{-3}M = 5 \text{ m}M$$

The turnover number, $k_{cat} = V_{max}/[E]_T$. Therefore, **one must know $[E]_T$ in addition to V_{max} to calculate the turnover number.**

(b)

$$[ES] = \frac{[E]_T [S]}{K_M + [S]}$$

$$\frac{[ES]}{[E]_T} = \frac{[S]}{K_M + [S]}$$

For no inhibitor: $\dfrac{[ES]}{[E]_T} = \dfrac{5 \text{ m}M}{(3.3 + 5) \text{ m}M} = \dfrac{5}{8.3} = 0.60$

For type (2) inhibitor: $\dfrac{[ES]}{[E]_T} = \dfrac{[S]}{\alpha K_M + [S]}$

$$\frac{[ES]}{[E]_T} = \frac{5 \text{ m}M}{(2.33 \times 3.3 + 5) \text{ m}M} = \frac{5}{7.7 + 5} = 0.39$$

For type (3) inhibitor: $\dfrac{[ES]}{[E]_T} = \dfrac{[S]}{\alpha K_M + \alpha'[S]}$

But since $\alpha = \alpha'$ in this case,

$$\frac{[ES]}{[E]_T} = \frac{[S]}{\alpha'(K_M + [S])} = \frac{5\,mM}{3.00(3.3 + 5)\,mM} = \frac{5}{24.9} = 0.20$$

5. Since the molecular mass of methanol (MeOH) is 32D, 100 mL of methanol in 40L of solution has the concentration

$$[MeOH] = \frac{100\,mL \times 0.79\,g{\bullet}mL^{-1}}{32\,g{\bullet}mol^{-1} \times 40\,L} = 0.062M$$

Ethanol (EtOH) is a competitive inhibitor of methanol with LADH.

$$v_o(\text{uninhibited}) = \frac{V_{max}[S]}{K_M + [S]}$$

$$v_o(\text{inhibited}) = \frac{V_{max}[S]}{\alpha K_M + [S]}$$

where $\alpha = \left(1 + \dfrac{[I]}{K_I}\right)$, $[MeOH] = [S]$, and $K_M = 0.01M$.

We require that

$$\frac{v_o(\text{inhibited})}{v_o(\text{uninhibited})} = 0.05 = \frac{K_M + [MeOH]}{\alpha K_M + [MeOH]}$$

$$0.05 = \frac{0.01 + 0.062}{0.01\alpha + 0.062}$$

$$0.0005\alpha + 0.0031 = 0.072$$

$$\alpha = 138 = \left(1 + \frac{[I]}{K_I}\right)$$

$$[I] = (138 - 1)K_I = 137K_I = 137 \times 1.0 \times 10^{-3}$$

$$[I] = 0.137M = [EtOH]$$

$$\text{Mol EtOH} = 0.137M \times 40\,L = 5.48\,mol$$

$$\text{Molecular mass EtOH} = 46\,D$$

$$\text{Mass EtOH} = 46 \text{ g} \bullet \text{mol}^{-1} \times 5.48 \text{ mol} = 252 \text{ g}$$

$$\text{Volume pure EtOH} = \frac{252 \text{ g}}{0.79 \text{ g} \bullet \text{mL}^{-1}} = 319 \text{ mL}$$

Since 100 proof whiskey is 50% by volume, it is necessary to imbibe 2 x 319 mL = **638 mL** (which is about 5/6 of a "fifth" of whiskey).

6. (a)

$$v_o = \frac{V_{max}[S]}{K_M + [S]}$$

$$0.43 \times 10^{-6} M \bullet \text{min}^{-1} = \frac{V_{max} \times 0.20M}{1.0 \times 10^{-4} M + 0.20M}$$

$$V_{max} = 0.43 \times 10^{-6} M \bullet \text{min}^{-1}$$

so at [S] = 0.02M,

$$v_o = \frac{0.43 \times 10^{-6} M \bullet \text{min}^{-1} \times 0.02M}{1.0 \times 10^{-4} M + 0.02M} = 0.43 \times 10^{-6} M \bullet \text{min}^{-1}$$

(b) The range near K_M, that is, [S] $\approx$ 1.0 x 10^{-4}M since this is the range that the denominator of the Michaelis-Menten equation varies with both K_M and [S].

7. The effects of competitive inhibitors can be diluted out by substrate whereas those of uncompetitive and mixed inhibitors cannot.

8. Other kinetic models, perhaps more complex and implausible than that under consideration, can be constructed which are described by equations of identical form. Consequently, kinetic measurements can be used to reject a particular kinetic model, but not to prove one.

9.

$$V'_{max} = V_{max}/f_2 \text{ where } f_2 = \frac{[H^+]}{K_{ES1}} + 1 + \frac{K_{ES2}}{[H^+]}$$

Differentiating to find the value of [H$^+$] at which V'_{max} is maximal,

$$\frac{dV'_{max}}{d[H^+]} = \frac{-V_{max}}{f_2^2} \left(\frac{1}{K_{ES1}} + \frac{-K_{ES2}}{[H^+]_{max}^2} \right) = 0$$

$$\frac{1}{K_{ES1}} + \frac{-K_{ES2}}{[H^+]^2_{max}} = 0$$

$$[H^+]_{max} = \sqrt{K_{ES1}K_{ES2}}$$

For $pK_{ES1} = 4$ and $pK_{ES2} = 8$, $K_{ES1} = 10^{-4}M$ and $K_{ES2} = 10^{-8}M$.

Then, $[H^+]_{max} = \sqrt{10^{-4}M \times 10^{-8}M} = 10^{-6}M$ so that $pH_{max} = 6.$

At this pH,

$$\frac{V'_{max}}{V_{max}} = \frac{1}{f_2} = \frac{1}{\dfrac{[H^+]}{K_{ES1}} + 1 + \dfrac{K_{ES2}}{[H^+]}} = \frac{1}{\dfrac{10^{-6}M}{10^{-4}M} + 1 + \dfrac{10^{-8}M}{10^{-6}M}}$$

$$= \frac{1}{10^{-2} + 1 + 10^{-2}} = \frac{1}{1.02} = 0.98$$

10.

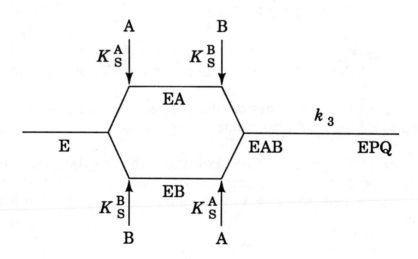

$$K_S^A = \frac{[E][A]}{[EA]} = \frac{[EB][A]}{[EAB]} \qquad K_S^B = \frac{[E][B]}{[EB]} = \frac{[EA][B]}{[EAB]}$$

$$v_0 = k_3[EAB]$$

Initially: $[E]_T = [E] + [EA] + [EB] + [EAB]$

$$[EA] = \left(\frac{K_S^B}{[B]}\right)[EAB]$$

$$[EB] = \left(\frac{K_S^A}{[A]}\right)[EAB]$$

$$[E] = \frac{K_S^A [EA]}{[A]} = \left(\frac{K_S^A K_S^B}{[A][B]}\right)[EAB]$$

$$[E]_T = \left(\frac{K_S^A K_S^B}{[A][B]} + \frac{K_S^B}{[B]} + \frac{K_S^A}{[A]} + 1\right)[EAB]$$

$$v_0 = k_3 [EAB] = \frac{k_3 [E]_T}{\left(\dfrac{K_S^A K_S^B}{[A][B]} + \dfrac{K_S^B}{[B]} + \dfrac{K_S^A}{[A]} + 1\right)} = \frac{\dfrac{V_{max}[A][B]}{K_S^A K_S^B}}{1 + \dfrac{[A]}{K_S^A} + \dfrac{[B]}{K_S^B} + \dfrac{[A][B]}{K_S^A K_S^B}}$$

where $V_{max} = k_3 [E]_T$.

11. (a) Conservation: $\quad [E]_T = [E] + [EA] + [F] + [FB]$

Equilibrium: $\quad K_S^A = \dfrac{[E][A]}{[EA]}, \quad K_S^B = \dfrac{[F][B]}{[FB]}$

$$[E]_T = \left(\frac{K_S^A}{[A]} + 1\right)[EA] + \left(\frac{K_S^B}{[B]} + 1\right)[FB]$$

$$[FB] = \frac{[E]_T - \left(\dfrac{K_S^A}{[A]} + 1\right)[EA]}{\left(\dfrac{K_S^B}{[B]} + 1\right)}$$

Steady state: At the steady state, the concentrations of the intermediates remain constant and the flows of reactants through each stage of the reaction are equal, that is,

$$\frac{d[F]}{dt} = \frac{d[FB]}{dt} = k_2 [EA] - k_4 [FB] = 0$$

so that

$$[EA] = \left(\frac{k_4}{k_2}\right)[FB]$$

$$[FB] = \cfrac{[E]_T - \left(\dfrac{K_S^A}{[A]} + 1\right)\dfrac{k_4}{k_2}[FB]}{\left(\dfrac{K_S^B}{[B]} + 1\right)}$$

$$[FB] = \cfrac{[E]_T}{\left(\dfrac{K_S^A}{[A]} + 1\right)\dfrac{k_4}{k_2} + \left(\dfrac{K_S^B}{[B]} + 1\right)}$$

But $v_0 = k_4\,[FB]$ and $V_{max} = k_4\,[E]_T$ so that

$$v_0 = \cfrac{V_{max}}{\left(\dfrac{K_S^A}{[A]} + 1\right)\dfrac{k_4}{k_2} + \left(\dfrac{K_S^B}{[B]} + 1\right)}$$

(b)

$$\frac{1}{v_0} = \left(\frac{k_4}{k_2}\right)\left(\frac{K_S^A}{V_{max}}\right)\frac{1}{[A]} + \left(\frac{K_S^B}{V_{max}}\right)\frac{1}{[B]} + \left(\cfrac{1 + \dfrac{k_4}{k_2}}{V_{max}}\right)$$

Lineweaver-Burk plots of $1/v_0$ vs $1/[A]$ at constant $[B]$ are a **series of parallel lines.**

(c)   Lineweaver-Burk plots of $1/v_0$ vs $1/[B]$ at constant $[A]$ are a **series of parallel lines.**

12. The Lineweaver-Burk plots in the absence of products for either substrate gives a mixed inhibition-type pattern. This is consistent with a sequential mechanism. The product inhibition pattern is:

Product Inhibitor	[MgATP] variable	[phosphocreatine] variable
MgATP^{2-}	Competitive	Competitive
Creatine	Competitive	Competitive

This pattern is consistent with a **Rapid Equilibrium Random Bi Bi mechanism** (Table 13-3).

Chapter 14
ENZYMATIC CATALYSIS

---

1. γ-Pyridone is subject to the same sort of tautomerization as is α-pyridone:

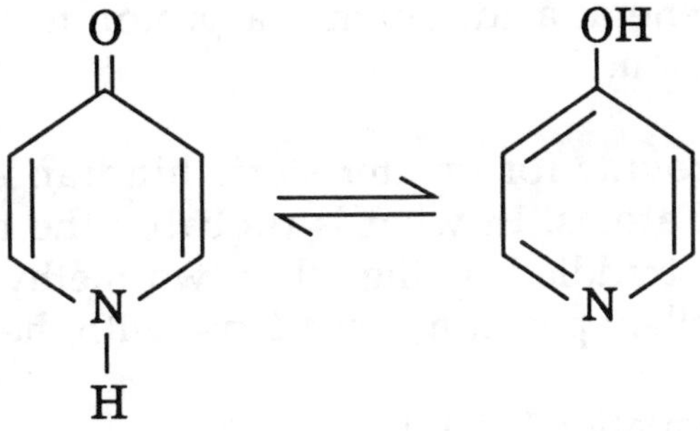

but its acidic and basic groups are geometrically unable to simultaneously interact with glucose. β-Pyridone does not exist; only β-hydroxypyridine:

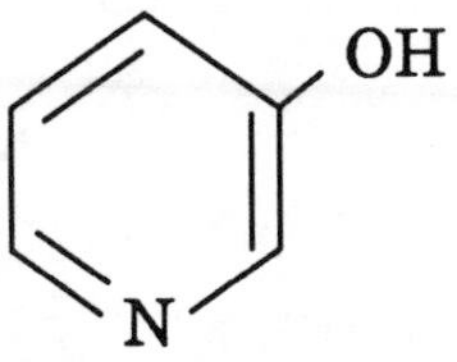

Even so, its acid–base groups cannot simultaneously interact with glucose.

2. The mechanism of base-catalyzed hydrolysis of RNA resembles the RNase catalyzed reaction

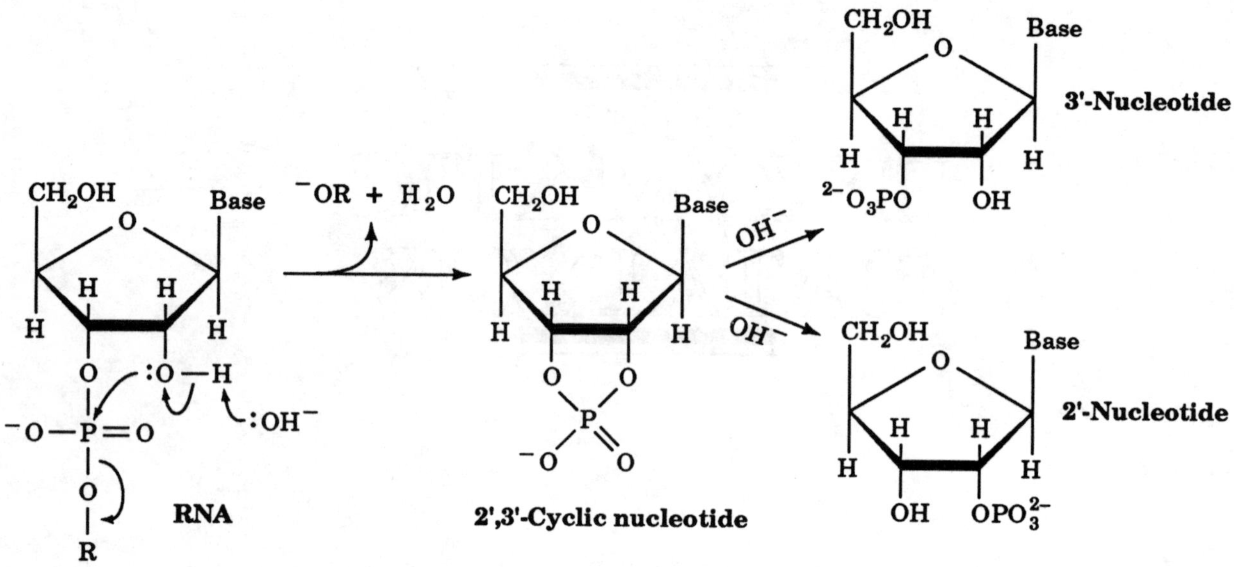

DNA is not subject to this mechanism because it lacks the 2′-OH group to make the initial nucleophilic attack on the phosphate group.

3. In the first step of the carboxypeptidase reaction, the Zn^{2+} ion activates its bound water as a nucleophile by making it more acidic (metal ion catalysis). Glu 270 aids this process by abstracting a proton from the water (general base catalysis). In the second reaction step, Glu 270, acting as a general acid, donates a proton to the amine leaving group thereby promoting bond scission.

4. The three methyl groups provide for greater steric hindrance to rotation about their connecting bond than do H atoms. In what is probably the most stable position, the methyl group in the ring is straddled by the other two methyl groups thereby placing the carboxyl group in an excellent position to condense with the hydroxyl group.

5. Following the derivation of equation [14-11]:

$$\text{Uncatalyzed reaction:} \quad A + B \xrightarrow{\ k_N\ } P$$

$$\text{Enzymatic reaction:} \quad EAB \xrightarrow{\ k_E\ } EP$$

The two reactions are related by the following scheme:

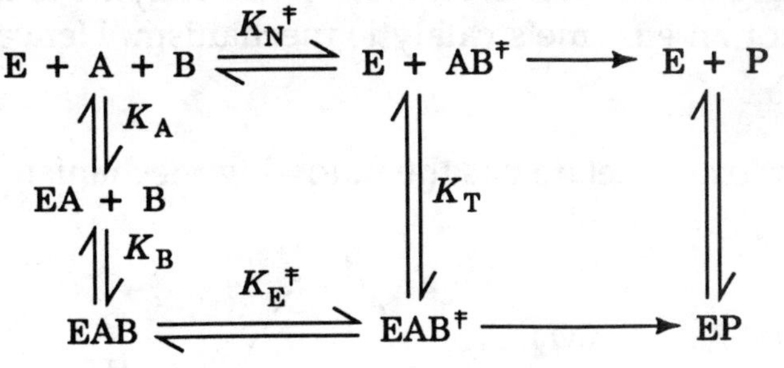

$$E + A + B \underset{}{\overset{K_N^{\ddagger}}{\rightleftharpoons}} E + AB^{\ddagger} \longrightarrow E + P$$

with vertical equilibria K_A, K_T; then

$$EA + B$$

$$\downarrow K_B \qquad K_E^{\ddagger}$$

$$EAB \rightleftharpoons EAB^{\ddagger} \longrightarrow EP$$

where

$$K_A = \frac{[EA]}{[E][A]}, \quad K_B = \frac{[EAB]}{[EA][B]}, \quad K_T = \frac{[EAB^{\ddagger}]}{[E][AB^{\ddagger}]}$$

$$K_N^{\ddagger} = \frac{[E][AB^{\ddagger}]}{[E][A][B]}, \quad K_E^{\ddagger} = \frac{[EAB^{\ddagger}]}{[EAB]}$$

Thus,

$$\frac{K_T}{K_A K_B} = \frac{[A][B][EAB^{\ddagger}]}{[AB^{\ddagger}][EAB]} = \frac{K_E^{\ddagger}}{K_N^{\ddagger}}$$

Now

$$v_N = k_N [A][B] = \left(\frac{\kappa k_B T}{h}\right)[AB^{\ddagger}] = \left(\frac{\kappa k_B T}{h}\right) K_N^{\ddagger} [A][B]$$

and

$$v_E = k_E [EAB] = \left(\frac{\kappa k_B T}{h}\right)[EAB^{\ddagger}] = \left(\frac{\kappa k_B T}{h}\right) K_E^{\ddagger} [EAB]$$

Therefore,

$$\frac{k_E}{k_N} = \frac{K_E^{\ddagger}}{K_N^{\ddagger}} = \frac{K_T}{K_A K_B}$$

6. Two such analogs are:

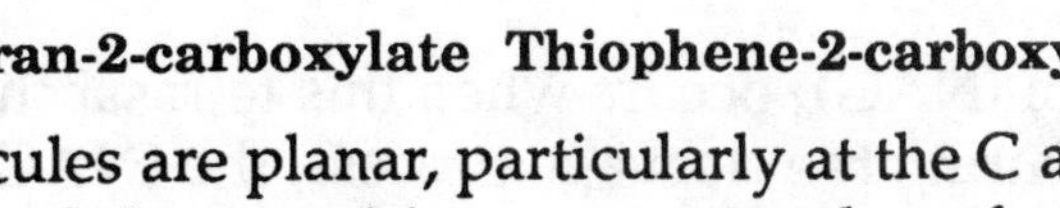

Furan-2-carboxylate Thiophene-2-carboxylate

Both of these molecules are planar, particularly at the C atom to which the carboxylate is bonded, as is true of the transition state complex of proline in the proline racemase reaction.

7. The preferential binding of the transition state to an enzyme is an important (often the most important) part of an enzyme's catalytic mechanism. Hence, the substrate binding site is the catalytic site.

8. The decarboxylation of oxaloacetate has the following mechanism:

Enolpyruvate intermediate

Pyruvate

Oxalate,

resembles this reaction's enolpyruvate intermediate and hence efficiently inhibits the enzyme.

9. Enzymes must be able to bind substrates and hold catalytic groups in the proper orientations for reaction; they must bind enough of their substrates to be able to preferentially bind the transition state complex; and they must provide the proper electric field for electrostatic catalysis. These extensive requirements can only be met by a large complex molecule. A protrusion, rather than a cleft, would be unable to provide any of the above environmental effects. Moreover, it seems unlikely that a protrusion would have sufficient structural complexity or rigidity to properly orient its catalytic groups for efficient catalysis.

10. Phe 34, Ser 36, and Trp 108 are all in the vicinity of Glu 35 which is the proton donor in the lysozyme reaction. Glu 35 can act in this manner because its location in a predominantly nonpolar region of the protein gives it an abnormally high pK. Changing the above residues to the positively charged Arg residue should significantly lower the pK of Glu 35. Consequently, such a lysozyme variant should have greatly reduced pH range of enzymatic activity because its Glu 35 is incapable of acting as a proton donor above its pK.

11. The productive cleavage of $(NAG)_4$ occurs when this tetrasaccharide binds in the C, D, E and F sites of lysozyme. During the cleavage process, the $(NAG)_2$ leaving group that had bound in the E and F sites drifts away and is normally replaced by OH^- to complete the hydrolytic reaction. If, however, the leaving group is instead replaced by $(NAG)_4$

binding such that its first two residues occupy the E and F sites, then the lysozyme reaction can proceed in reverse to yield $(NAG)_6$.

$$(NAG)_4 \longrightarrow (NAG)_2 \longrightarrow (NAG)_6$$
$$\searrow \qquad \nearrow$$
$$(NAG)_2 \qquad (NAG)_4$$

Here the $(NAG)_4$ rather than OH^- acts as the acceptor in the reaction (such a process is known as transglycosylation). The transglycosylation reaction is indicative of the existence of the glycosyl-enzyme intermediate that is postulated to occur in the Phillips mechanism.

12. The double bond in the last ring of the inhibitor makes its C(1) atom trigonal thereby causing this ring to assume the half-chair conformation of the transition state complex in the lysozyme reaction. This substance is therefore a transition state analog inhibitor of lysozyme.

13. The positively charged and small nonpolar residues for which trypsin and elastase are specific are more likely to be on the surface of a protein than the bulky nonpolar residues for which chymotrypsin is specific. The autolysis of chymotrypsin is, consequently, slower than that of trypsin and elastase.

14. The observation that subtilisin and chymotrypsin are genetically unrelated indicates that their active site geometries arose by convergent evolution. Assuming that evolution has optimized the catalytic efficiencies of these enzymes and that there is only one optimal arrangement of catalytic groups, any similarities between the active sites of subtilisin and chymotrypsin must be of catalytic significance. Conversely, any differences are unlikely to be catalytically important.

15. Benzamidine, being a cationic and reasonably bulky molecule, binds in the specificity pocket of trypsin thereby blocking true substrates from binding to the enzyme. Leupeptin resembles a polypeptide. Its cationic guanidino side chain binds in the specificity pocket of trypsin and its C-terminal aldehyde group occupies the position normally held by the carbonyl group of the scissile peptide. This aldehyde group reacts, as does a normal substrate, with Ser 195 to form a tetrahedral adduct:

However, the catalytic reaction cannot progress beyond this point since the resulting hemiacetal ion lacks the proper leaving group.

Such transition state analogs will inhibit chymotrypsin and elastase if they have the proper terminal side chain to bind in the specificity pocket of their respective enzymes. Chymotrypsin requires a bulky aromatic group to bind in its specificity pocket whereas elastase preferentially binds small nonpolar side chains such as CH_3. Thus, examples of leupeptinlike chymotrypsin and elastase inhibitors are:

$$CH_3-\overset{\overset{O}{\|}}{C}-Leu-Leu-NH-\overset{}{\underset{\underset{R}{|}}{C}H}-\overset{\overset{O}{\|}}{C}H$$

$R = -CH_2-\langle\ \rangle$ **Chymotrypsin inhibitor**

$R = -CH_3$ **Elastase inhibitor**

16. Since boronic acids easily form tetrahedral adducts and chymotrypsin and subtilisin preferentially bind bulky nonpolar side chains such as a phenyl ring, phenylethyl boronic acid readily reacts with these enzymes to form a covalently bound transition state analog of the enzymatic reaction intermediate:

$$H_2C-\overset{\overset{OH}{|}}{\underset{\underset{CH_2}{|}}{B}}-OH$$
$$O-CH_2-Ser\ 195$$

Enzyme
specificity pocket

17. If the soybean trypsin inhibitor was not removed from tofu, it would inhibit the trypsin in the intestine. At best, this would reduce the nutritional value of the meal by rendering its protein indigestible. It might very well also lead to intestinal upset.

18. The peptide bond that is cleaved to activate both trypsinogen and chymotrypsingen is preceeded by a cationic residue for which trypsin but not chymotrypsin is specific. Thus, trypsin, but not chymotrypsin, can activate both of these zymogens.

19. The only other readily reduced group in glutathione reductase besides the redox-active disulfide is its FAD prosthetic group. Hence, the 4-electron reduced enzyme is EH_2 with its FAD reduced to $FADH_2$ and its cleaved (reduced) disulfide group in complex with arsenite.

20. Iodoacetate reacts specifically with thiol groups (Section 5-1B):

$$RSH \ + \ ICH_2COO^- \longrightarrow R-S-CH_2COO^- \ + \ HI$$

Thus, in EH_2, iodoacetate reacts with Cys 58. This inactivates the enzyme since this blocked Cys 58 can neither react with GSSR in a disulfide exchange reaction nor back react to reform the redox-active disulfide that normally reduces $NADP^+$.

Chapter 15

INTRODUCTION TO METABOLISM

1. Overall stoichiometry: **2ATP + 2H$_2$O $\rightleftharpoons$ 2ADP + 2P$_i$**. These pathways must be independently controlled or the ATP formed in glycolysis would be automatically used in gluconeogenesis thereby uselessly dissipating the ATP formed. Conversely, any glucose formed by gluconeogenesis would be rapidly broken down again by glycolysis. These pathways cannot be separately controlled unless they differ in at least one reaction. This is because altering the activity of an enzyme changes the rate of the forward reaction by the same proportion as that of the back reaction. Moreover, since glycolysis is exergonic, at least one reaction of gluconeogenesis must differ from that of glycolysis so that free energy can be supplied to make gluconeogenesis also exergonic.

2.

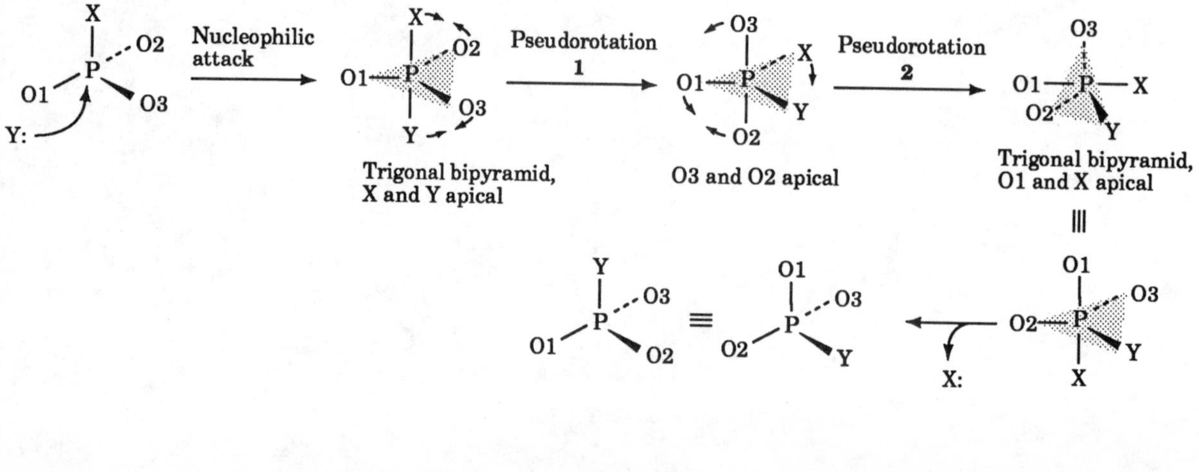

Nucleophilic attack
Pseudorotation 1
Pseudorotation 2
Trigonal bipyramid, X and Y apical
O3 and O2 apical
Trigonal bipyramid, O1 and X apical

Comparison of the first and last structures indicates that the nucleophilic displacement reaction in which 2 pseudorotations have taken place occurs with inversion of configuration (as it does without the pseudorotations).

3. For ^{14}C, $t_{1/2}$ = 5730 years (Table 15-1) so that the first-order rate constant for its disintegration is:

$$k = \frac{0.693}{t_{1/2}} = \frac{0.693}{5730 \text{ year}} = 1.21 \times 10^{-4} \text{ y}^{-1}$$

$$= 1.21 \times 10^{-4} \text{ y}^{-1} \times \left(\frac{1 \text{ y}}{365 \text{ day}}\right) \times \left(\frac{1 \text{ day}}{24 \text{ h}}\right) \times \left(\frac{1 \text{ h}}{3600 \text{ s}}\right) = 3.84 \times 10^{-12} \text{ s}^{-1}$$

Thus, one mole of ^{14}C should have $6.02 \times 10^{23} \times 3.84 \times 10^{-12}$ s^{-1} = 2.31×10^{12} disintegrations/s.

A sample with 5 μCi/μmol has 5 Ci/mol = $5 \times 3.70 \times 10^{10} = 1.85 \times 10^{11}$ disintegrations/s. Its C atoms are therefore $(1.85 \times 10^{11}/2.31 \times 10^{12}) \times 100 = $ **8% ^{14}C.**

4. Subtracting reaction (2) from reaction (1) yields the ATP hydrolysis reaction:

$$ATP + H_2O \rightleftharpoons ADP + P_i$$

The ΔG^{o}'s of the component reactions, being state functions, subtract in the same way. Hence,

$$\Delta G^{o'}(ATP) = \Delta G_1^{o'} - \Delta G_2^{o'} = -16.3 - 14.2 = \textbf{-30.5 kJ•mol}^{-1}$$

5. From Table 15-3,

$$ATP + H_2O \rightleftharpoons ADP + P_i \quad \Delta G^{o'} = -30.5 \text{ kJ•mol}^{-1}$$

$$\text{Glucose-6-P} + H_2O \rightleftharpoons \text{Glucose} + P_i \quad \Delta G^{o'} = -13.8 \text{ kJ•mol}^{-1}$$

Subtracting the two reactions,

$$ATP + \text{Glucose} \rightleftharpoons \text{Glucose-6-P} + ADP \quad \Delta G^{o'} = -30.5 + 13.8$$
$$= -16.7 \text{ kJ•mol}^{-1}$$

$$K_{eq}' = \frac{[G6P][ADP]}{[G][ATP]} = e^{-\Delta G^{o'}/RT} = e^{16,700/8.314 \times 298} = 846$$

Conservation:

$$[ATP] + [ADP] = 40 \text{ m}M = 0.040M$$

$$[G6P] + [G] = 20 \text{ m}M = 0.020M$$

Stoichiometry:

$$[G6P] = [ADP] = x$$

$$\frac{x^2}{(0.020 - x)(0.040 - x)} = 846$$

$$x^2 = 846x^2 - 50.8x + 0.677$$

$$845x^2 - 50.8x + 0.677 = 0$$

Using the quadratic equation,

$$x = \frac{50.8 \pm \sqrt{50.8^2 - 4 \times 845 \times 0.677}}{2 \times 845} = \frac{50.8 \pm 17.1}{1690}$$

$$x = 0.04018M \text{ or } 0.01994M$$

The former root is physically unrealistic as it is larger than the maximum possible, 0.040M. Hence,

$$[ADP] = [G6P] = \mathbf{0.01994\textit{M}}$$

$$[G] = 0.02000 - 0.01994 = \mathbf{0.00006\textit{M}}$$

$$[ATP] = 0.04000 - 0.01994 = \mathbf{0.02006\textit{M}}$$

6. Free energy captured as ATP = $-50 \text{ kJ} \bullet \text{mol}^{-1} \times 38 = -1900 \text{ kJ} \bullet \text{mol}^{-1}$

Free energy liberated upon glucose combustion (from Problem 3-8a) = $-2823.2 \text{ kJ} \bullet \text{mol}^{-1}$

$$\text{Efficiency} = \frac{-1900 \text{ kJ} \bullet \text{mol}^{-1} \times 100}{-2823 \text{ kJ} \bullet \text{mol}^{-1}} = \mathbf{67.3\%}$$

7. (a) $\qquad \text{ATP} + \text{H}_2\text{O} \rightleftharpoons \text{ADP} + \text{P}_i \qquad \Delta G^{\circ\prime} = -30.5 \text{ kJ} \bullet \text{mol}^{-1}$

$$\Delta G = \Delta G^{\circ\prime} + RT \ln Q' = \Delta G^{\circ\prime} + RT \ln \frac{[\text{ADP}][\text{P}_i]}{[\text{ATP}]}$$

$$= -30.5 + \frac{8.314 \times 298}{1000} \ln \left(\frac{5 \times 10^{-4} \times 10^{-3}}{5 \times 10^{-3}} \right) = -30.5 - 22.8 = \mathbf{-53.3 \text{ kJ} \bullet \text{mol}^{-1}}$$

(b)

$$ATP + H_2O \rightleftharpoons ADP + P_i \qquad\qquad \Delta G^{o'} = -30.5 \text{ kJ} \cdot \text{mol}^{-1}$$

$$Creatine + P_i \rightleftharpoons Phosphocreatine \qquad \Delta G^{o'} = 43.1 \text{ kJ} \cdot \text{mol}^{-1}$$

$$ATP + Creatine \rightleftharpoons ATP + Phosphocreatine \quad \Delta G^{o'} = 12.6 \text{ kJ} \cdot \text{mol}^{-1}$$

$$K'_{eq} = \frac{[phosphocreatine][ADP]}{[creatine][ATP]} = e^{-\Delta G^{o'}/RT} = e^{-12.6 \times 1000/8.314 \times 298} = 0.0062$$

$$\frac{[phosphocreatine]}{[creatine]} = \frac{[ATP]}{[ADP]} \times 0.0062 = \frac{5 \times 10^{-3}}{0.5 \times 10^{-3}} \times 0.0062 = \mathbf{0.062}$$

(c) From Part (b):

$$\frac{[phosphocreatine]}{[creatine]} = \frac{[ATP]}{[ADP]} \times 0.0062 = 1$$

$$\frac{[ATP]}{[ADP]} = \frac{1}{0.0062} = \mathbf{161}$$

$$\Delta G = \Delta G^{o'} + RT \ln \frac{[ADP][P_i]}{[ATP]}$$

$$= -30.5 + \frac{8.314 \times 298}{1000} \ln \frac{10^{-3}}{161}$$

$$\Delta G = -30.5 - 29.7 = \mathbf{-60.2 \text{ kJ} \cdot \text{mol}^{-1}}$$

8. (a) From Table 15-3:

$$ATP + H_2O \rightleftharpoons AMP + PP_i \qquad \Delta G^{o'} = -32.2 \text{ kJ} \cdot \text{mol}^{-1}$$

$$PP_i + H_2O \rightleftharpoons 2P_i \qquad\qquad \Delta G^{o'} = -33.5 \text{ kJ} \cdot \text{mol}^{-1}$$

$$2ADP \rightleftharpoons ATP + AMP \qquad\quad \Delta G^{o'} = -4.7 \text{ kJ} \cdot \text{mol}^{-1}$$

$$K'_{eq} = \frac{[ATP][AMP]}{[ADP]^2} = e^{-\Delta G^{o'}/RT}$$

$$[AMP] = \frac{[ADP]^2}{[ATP]} e^{-\Delta G^{\circ\prime}/RT} = \frac{(5 \times 10^{-4})^2}{5 \times 10^{-3}} e^{-(-4700)/8.314 \times 298}$$

$$[AMP] = 3.33 \times 10^{-4}M = \textbf{0.33 m}\boldsymbol{M}$$

(b) For the hydrolysis of ATP to ADP + P_i,

$$Q = \frac{[ADP][P_i]}{[ADP]} = e^{(\Delta G - \Delta G^{\circ})/RT}$$

$$\frac{[ADP]}{[ATP]} = \frac{1}{[P_i]} e^{(\Delta G - \Delta G^{\circ\prime})/RT} = \frac{1}{10^{-3}} e^{(-55 + 30.5) \times 1000/8.314 \times 298} = 0.051$$

$$[ATP] + [ADP] = 5 \times 10^{-3}\ M + 5 \times 10^{-4}\ M = 5.5 \times 10^{-3}M$$

$$[ADP]\,(1/0.051 + 1) = 5.5 \times 10^{-3}M$$

$$[ADP] = \frac{5.5 \times 10^{-3}M}{20.6} = 2.7 \times 10^{-4}\ M$$

$$[AMP] = \frac{[ADP]^2}{[ATP]} e^{-\Delta G^{\circ\prime}/RT} = \left(\frac{[ADP]}{[ATP]}\right)[ADP]\, e^{-(-4700)/8.314 \times 298}$$

$$[AMP] = 0.051 \times 2.7 \times 10^{-4}M \times 6.66 = \textbf{9.2} \times \textbf{10}^{-5}\boldsymbol{M} = \textbf{0.092 m}\boldsymbol{M}$$

9. The larger the reducing potential, the greater the reducing power. Therefore, according to Table 15-4:

		$E^{\circ\prime}(V)$
(c)	O_2	0.815
(e)	Cytochrome c	0.254
(a)	Fumarate$^-$	0.031
(f)	Lipoic acid	−0.29
(d)	NADP$^+$	−0.320
(b)	Cystine	−0.340

10.

$$\text{Acetoacetate}^- + 2H^+ + 2e^- \rightleftharpoons \beta\text{-hydroxybutyrate}^- \qquad E^{\circ\prime} = -0.346\ V$$

$$\text{NADH} \rightleftharpoons NAD^+ + H^+ + 2e^- \qquad -E^{\circ\prime} = 0.315\ V$$

$$\overline{\text{Acetoacetate}^- + \text{NADH} + H^+ \rightleftharpoons \beta\text{-hydroxybutyrate}^- + NAD^+ \quad \Delta E^{\circ\prime} = -0.031\ V}$$

According to the Nernst equation, equilibrium is attained when ΔE vanishes. Therefore,

$$K'_{eq} = \frac{[\beta\text{-hydroxybutyrate}^-][NAD^+]}{[\text{acetoacetate}^-][NADH]} = e^{nF\Delta E^{\circ\prime}/RT}$$

$$= e^{2 \times 96,494\, J\bullet V^{-1}\, mol^{-1} \times (-0.031\ V)/8.314\, J\bullet mol^{-1}\ K^{-1} \times 298\ K} = 0.09$$

Conservation indicates that

$$[\text{acetoacetate}^-] + [\beta\text{-hydroxybutyrate}^-] = 0.01M$$

$$[NADH] + [NAD^+] = 0.005M$$

Stoichiometry indicates that

$$[\beta\text{-hydroxybutyrate}^-] = [NAD^+] = x$$

Therefore,

$$[\text{acetoacetate}^-] = 0.01 - x$$

$$[NADH] = 0.005 - x$$

$$\frac{x^2}{(0.01 - x)(0.005 - x)} = 0.09$$

Using the quadratic equation:

$$x = \frac{-1.35 \times 10^{-3} \pm \sqrt{(1.35 \times 10^{-3})^2 - 4 \times 0.91 \times (-4.5 \times 10^{-6})}}{2 \times 0.91}$$

$$x = \frac{-1.35 \times 10^{-3} \pm 4.27 \times 10^{-3}}{1.82}$$

Choosing the positive root since this gives the only positive answer,

$$x = 0.0016M$$

$$[\beta\text{-hydroxybutyrate}^-] = [NAD^+] = \mathbf{0.0016}M$$

$$[\text{acetoacetate}^-] = 0.0100 - 0.0016 = \mathbf{0.0084}M$$

$$[\text{NADH}] = 0.0050 - 0.0016 = \mathbf{0.0034}M$$

11. $$\Delta G^{\circ\prime} = -nF\Delta E^{\circ\prime}$$

Therefore, the minimum redox potential required for ATP synthesis under standard conditions is

$$\Delta E^{\circ\prime} = \frac{-\Delta G^{\circ\prime}}{nF} = \frac{-30.5 \times 10^3 \text{ J} \bullet \text{mol}^{-1}}{2 \times 96,494 \text{ J} \bullet \text{V}^{-1} \text{mol}^{-1}} = 0.158 \text{ V}$$

From Table 15-4:

(a)

$$\text{NO}_3^- + 2\text{H}^+ + 2e^- \rightleftharpoons \text{NO}_2^- + \text{H}_2\text{O} \qquad\qquad E^{\circ\prime} = 0.42 \text{ V}$$

$$\text{Ethanol} \rightleftharpoons \text{Acetaldehyde} + 2\text{H}^+ + 2e^- \qquad\qquad -E^{\circ\prime} = 0.197 \text{ V}$$

$$\text{Ethanol} + \text{NO}_3^- \rightleftharpoons \text{Acetaldehyde} + \text{NO}_2^- + \text{H}_2\text{O} \qquad \Delta E^{\circ\prime} = 0.617 \text{ V}$$

$\Delta E^{\circ\prime} > 0.158$ V; **the nutrients are usable.**

(b)

$$\text{Fumarate}^- + 2\text{H}^+ + 2e^- \rightleftharpoons \text{Succinate}^- \qquad\qquad E^{\circ\prime} = 0.031 \text{ V}$$

$$\text{SO}_3^{2-} + \text{H}_2\text{O} \rightleftharpoons \text{SO}_4^{2-} + 2\text{H}^+ + 2e^- \qquad\qquad -E^{\circ} = -0.48 \text{ V}$$

$$\text{Fumarate}^- + \text{SO}_3^{2-} + \text{H}_2\text{O} \rightleftharpoons \text{Succinate}^- + \text{SO}_4^{2-} \qquad \Delta E^{\circ} = -0.449 \text{ V}$$

Equilibrium is far in the direction of reactants indicating that the nutrients have been used up. $\Delta E^{\circ\prime} < 0.158$ V; **starvation will occur.**

(c)

$$\text{S} + 2\text{H}^+ + 2e^- \rightleftharpoons \text{H}_2\text{S} \qquad\qquad E^{\circ} = -0.23 \text{ V}$$

$$2(\tfrac{1}{2}\text{H}_2 \rightleftharpoons \text{H}^+ + e^-) \qquad\qquad -E^{\circ\prime} = 0.42 \text{ V}$$

$$\text{H}_2 + \text{S} \rightleftharpoons \text{H}_2\text{S} \qquad\qquad \Delta E^{\circ} = 0.19 \text{ V}$$

$\Delta E^{\circ\prime} > 0.158$ V; **the nutrients are usable.**

(d)

$$\text{Acetaldehyde} + 2H^+ + 2e^- \rightleftharpoons \text{Ethanol} \qquad E^{o\prime} = -0.197 \text{ V}$$

$$\text{Acetaldehyde} + H_2O \rightleftharpoons \text{Acetate}^- + 3H^+ + 2e^- \qquad -E^{o\prime} = 0.581 \text{ V}$$

$$\overline{\text{2Acetyldehyde} + H_2O \rightleftharpoons \text{Ethanol} + \text{Acetate} + H^+ \quad \Delta E^{o\prime} = 0.384 \text{ V}}$$

$\Delta E^{o\prime} > 0.158$ V; **the nutrients are usable.**

12. (a)

$$\tfrac{1}{2}O_2 + 2H^+ + 2e^- \rightleftharpoons H_2O \qquad E^{o\prime} = 0.815 \text{ V}$$

$$2(\tfrac{1}{2}H_2 \rightleftharpoons H^+ + e^-) \qquad -E^{o\prime} = 0.42 \text{ V}$$

$$\overline{H_2 + \tfrac{1}{2}O_2 \rightleftharpoons H_2O \qquad\qquad \Delta E^{o\prime} = 1.235 \text{ V}}$$

$\Delta G^{o\prime} = -nF\,\Delta E^{o\prime} = -2 \times 96{,}494 \text{ J} \bullet \text{V}^{-1} \text{ mol}^{-1} \times 1.235 \text{ V} \times 10^{-3} \text{ kJ} \bullet \text{J}^{-1} = -238.3 \text{ kJ} \bullet \text{mol}^{-1}$

The reaction as written proceeds to the right since $\Delta G^{o\prime}$ is negative.

(b)

$$\text{NAD}^+ + H^+ + 2e^- \rightleftharpoons \text{NADH} \qquad E^{o\prime} = -0.315 \text{ V}$$

$$\text{Lactate}^- \rightleftharpoons \text{Pyruvate}^- + 2H^+ + 2e^- \qquad -E^{o\prime} = +0.185 \text{ V}$$

$$\overline{\text{NAD}^+ + \text{Lactate}^- \rightleftharpoons \text{Pyruvate}^- + \text{NADH} + H^+ \quad \Delta E^{o\prime} = -0.130 \text{ V}}$$

$\Delta G^{o\prime} = -2 \times 96{,}494 \text{ J} \bullet \text{V}^{-1} \text{ mol}^{-1} \times (0.130) \text{ V} \times 10^{-3} \text{ kJ} \bullet \text{J}^{-1} = 25.1 \text{ kJ} \bullet \text{mol}^{-1}$

The reaction as written proceeds to the left since $\Delta G^{o\prime}$ is positive.

13. For the ATP half-reaction: $\text{ADP} + P_i \rightleftharpoons \text{ATP}$

$$\Delta G_{\text{ATP}} = \Delta G^{o\prime} + RT\ln Q$$

$$= 30.5 \times 10^3 + RT\ln\frac{[\text{ATP}]}{[\text{ADP}][P_i]} = 30.5 \times 10^3 + 8.314 \times 298 \times \ln\left(\frac{1 \times 10^{-5}}{1 \times 10^{-2} \times 1 \times 10^{-2}}\right)$$

$$= 30{,}500 - 5700 = +24{,}800 \text{ J} \bullet \text{mol}^{-1}$$

For a pH gradient half-reaction: $2H^+(\text{low pH}) \rightleftharpoons 2H^+(\text{high pH})$

$$\Delta G_{pH} = 2RT\ln\frac{[H^+(\text{high pH})]}{[H^+(\text{low pH})]}$$

$$= 2 \times 2.303RT \{\log [H^+(\text{high pH})] - \log [H^+(\text{low pH})]\}$$

$$= -2 \times 2.303RT\Delta pH \qquad \text{where } \Delta pH = pH \text{ (high)} - pH \text{ (low)}$$

To get net synthesis:

$$\Delta G_{pH} + \Delta G_{ATP} < 0$$

$$\Delta G_{pH} < \Delta G_{ATP}$$

$$-2 \times 2.303RT\Delta pH < 24,800$$

Therefore,

$$\Delta pH > \frac{-24,800}{-2 \times 2.303RT} = \frac{24,800}{2 \times 2.303 \times 8.314 \times 298} = \textbf{2.17 pH units}$$

14.

$$\Delta G = \Delta G^{\circ\prime} + RT\ln\frac{[H^+(\text{gastric})]}{[H^+(\text{plasma})]} + RT\ln\frac{[Cl^-(\text{gastric})]}{[Cl^-(\text{plasma})]}$$

$\Delta G^{\circ\prime} = 0$ since reactants are identical to products.

$37°C = 313$ K.

$\Delta G = 40.1$ kJ•mol^{-1}

0.1 L of 0.15M HCl contains 0.015 mol HCl. Hence, the free energy required to produce 1L of gastric juice is 0.015 mol x 40.1 kJ•mol^{-1} = **0.6 kJ.**

Chapter 16
GLYCOLYSIS

1. See Figures 16-1 and 16-3.

2. The aldolase reaction is:

$$\text{FBP} \underset{\text{aldolase}}{\rightleftharpoons} \text{DHAP} + \text{GAP} \qquad K = \frac{[\text{DHAP}]\,[\text{GAP}]}{[\text{FBP}]}$$

$$\text{GAP} \underset{\text{TIM}}{\rightleftharpoons} \text{DHAP} \qquad K' = \frac{[\text{DHAP}]}{[\text{GAP}]} = 5.5$$

$$\frac{[\text{FBP}]}{[\text{GAP}]} = \frac{[\text{DHAP}]}{K} = \left(\frac{5.5}{K}\right)[\text{GAP}]$$

But, according to equation [3.16],

$$\Delta G^{\circ\prime} = -RT \ln K = 22.8 \text{ kJ} \bullet \text{mol}^{-1}$$

$$K = e^{-22.8 \times 10^3/RT} = e^{-22{,}800 \text{ J} \bullet \text{mol}^{-1}/8.314 \text{ J} \bullet \text{mol}^{-1}\text{K}^{-1} \times (37+273)\text{ K}} = e^{-8.85} = 1.43 \times 10^{-4} M$$

so that

$$\frac{[\text{FBP}]}{[\text{GAP}]} = \left(\frac{5.5}{1.43 \times 10^{-4} M}\right)[\text{GAP}] = 3.85 \times 10^4 M^{-1}\,[\text{GAP}]$$

(a)   When $[GAP] = 2 \times 10^{-5}M$, $\dfrac{[FBP]}{[GAP]} = 3.85 \times 10^{4}M^{-1} \times 2 \times 10^{-5}M = \mathbf{0.77}$

(b)   When $[GAP] = 10^{-3}M$, $\dfrac{[FBP]}{[GAP]} = 3.85 \times 10^{4}M^{-1} \times 10^{-3}M = \mathbf{38.5}$

3. The PFK reaction, with $\Delta G = -25.9$ kJ•mol^{-1} (Table 16-2) is highly exergonic and therefore essentially irreversible. Thus, with a blockage of the GAPDH reaction, FBP, DHAP and GAP will accumulate and distribute themselves according to their equilibrium ratios. Since $\Delta G^{\circ\prime}$ for the aldolase reaction is $+22.8$ kJ•mol^{-1} whereas $\Delta G^{\circ\prime}$ for the TIM reaction is $+7.9$ kJ•mol^{-1}, the equilibrium ratios of FBP, DHAP and GAP will greatly favor FBP.

4. (a)   Glucose $+$ 2NAD^{+} $+$ 2P$_i$ $+$ 2ADP $\longrightarrow$ 2Pyruvate $+$ 2NADH $+$ 2ATP $+$ 2H$_2$O

(b)   Glucose $+$ 2NAD^{+} $+$ 2AsO$_4^{3-}$ $+$ 2ADP $\longrightarrow$ 2Pyruvate $+$ 2NADH $+$ 2ADP–AsO$_3$ $+$ 2H$_2$O

$\qquad\qquad$ 2ADP–AsO$_3$ $+$ 2H$_2$O $\longrightarrow$ 2ADP $+$ 2AsO$_4^{3-}$

$\rule{12cm}{0.4pt}$

$\qquad$ Overall: Glucose $+$ 2NAD^{+} $\longrightarrow$ 2Pyruvate $+$ 2NADH

(c)   Arsenate is a poison because it uncouples ATP generation from glycolysis. Consquently, glycolytic energy generation cannot occur.

5. (a)   The electron donor is GAP whose aldehyde group becomes oxidized to a carboxyl group (1,3-BPG). The electron acceptor is pyruvate whose ketone group is reduced to a secondary alcohol (lactate).

(b)   The electron donor is GAP, as in Part (a). The electron acceptor is acetaldehyde which is reduced to ethanol.

6. (a)   Following C(1) through Figure 16-3, C(1) of glucose becomes C(3) of GAP which becomes **C(3) of pyruvate.**

(b)   C(4) of glucose becomes C(1) of GAP which becomes **C(1) of pyruvate.**

7. The reaction proceeds, as does the class I aldolase reaction, through step 3 in Figure 16-9 yielding G3P and a C$_3$ enamine intermediate. The erythrose-4-phosphate then reacts with the enamine as follows.

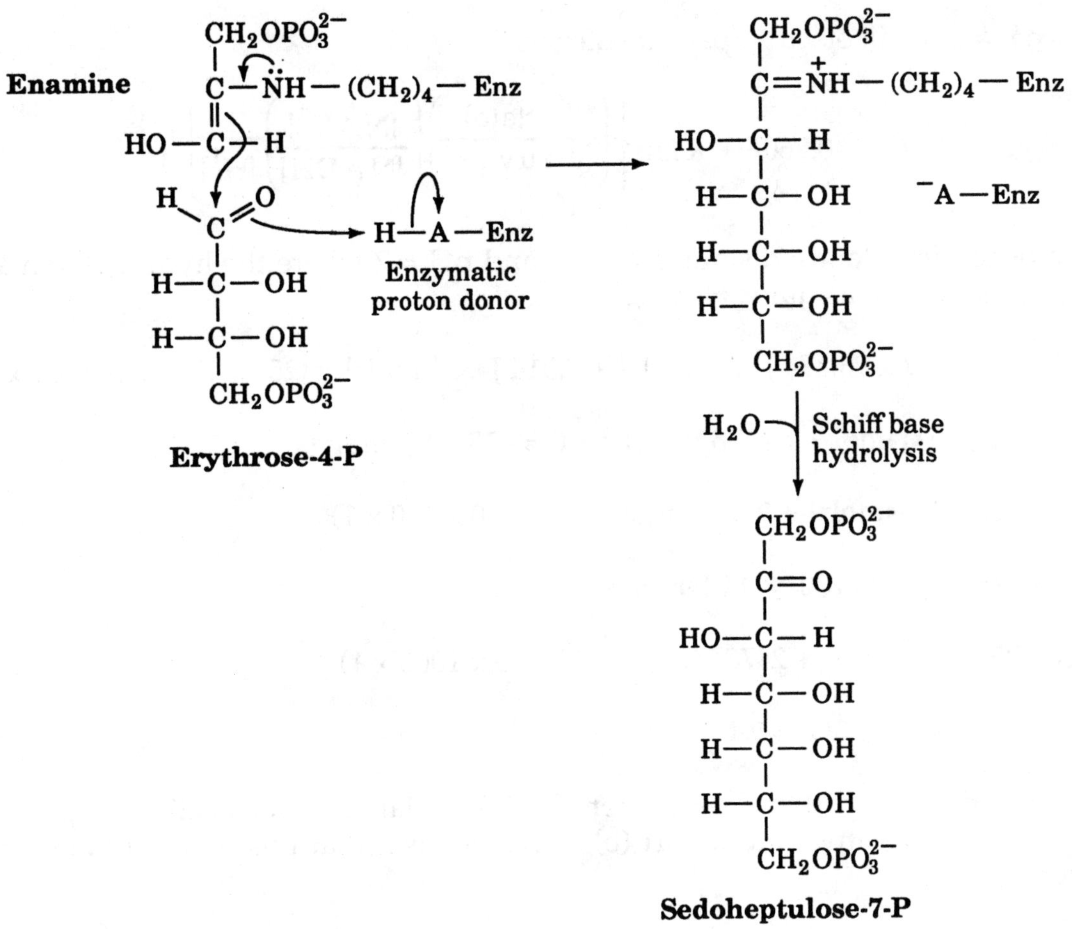

Enamine

Erythrose-4-P

Sedoheptulose-7-P

8.
$$\text{Pyruvate} + \text{NADH} + \text{H}^+ \rightleftharpoons \text{Lactate} + \text{NAD}^+$$

$$\Delta E = \Delta E^{\circ\prime} - \frac{RT}{nF} \ln \frac{[\text{lactate}][\text{NAD}^+]}{[\text{pyruvate}][\text{NADH}][\text{H}^+]}$$

$$\Delta E^{\circ\prime} = E^{\circ\prime}(e^- \text{ acceptor}) - E^{\circ\prime}(e^- \text{ donor}) = E^{\circ\prime} (\text{pyruvate}) - E^{\circ\prime} (\text{NAD}^+)$$

$$\Delta E^{\circ\prime} = -0.185 - (-0.315) = 0.130 \text{ V}.$$

Thus,

$$\Delta E = 0.130 \text{ V} - \frac{RT}{nF} \ln \frac{[\text{lactate}][\text{NAD}^+]}{[\text{pyruvate}][\text{NADH}][\text{H}^+]}$$

But

$$\Delta G = -nF\Delta E$$

$$\Delta G = -nF \times 0.130 + RT \ln \frac{[\text{lactate}][\text{NAD}^+]}{[\text{pyruvate}][\text{NADH}][\text{H}^+]}$$

and since $n = 2$ for an electron pair transfer,

$$\Delta G = -0.260F + RT\ln\left\{\left(\frac{[\text{lactate}]}{[\text{pyruvate}]}\right)\left(\frac{[\text{NAD}^+]}{[\text{NADH}]}\right)\frac{1}{[\text{H}^+]}\right\}$$

At the biochemical standard state, $T = 25\,°C$ and pH = 7 where the hydrogen ion activity (here $[\text{H}^+]$) is defined as unity. Hence,

(a) $\Delta G = -0.260\text{ V} \times 96,494 \text{ J}\bullet\text{V}^{-1}\text{ mol}^{-1} + 8.314 \text{ J}\bullet\text{K}^{-1}\text{ mol}^{-1} \times (25 + 273)\text{ K ln}(1 \times 1 \times 1)$

   $= -25,100 \text{ J}\bullet\text{mol}^{-1} + 2478 \text{ J}\bullet\text{mol}^{-1} \times 0 = \textbf{-25.1 kJ}\bullet\textbf{mol}^{-1}$

(b) $\Delta G = -25,100 \text{ J}\bullet\text{mol}^{-1} + 2478 \text{ J}\bullet\text{mol}^{-1} \times \ln (160 \times 160 \times 1)$

   $= -25,100 + 25,100 = \textbf{0 kJ}\bullet\textbf{mol}^{-1}$

(c) $\Delta G = 25,100 \text{ J}\bullet\text{mol}^{-1} + 2478 \text{ J}\bullet\text{mol}^{-1} \ln (1000 \times 1000 \times 1)$

   $= -25,100 + 34,200 = \textbf{9.1 kJ}\bullet\textbf{mol}^{-1}$

(d) The reaction, as written above, is for NADH oxidation. This reaction is spontaneous so long as ΔG is negative. From Part (b), it can be seen that this occurs under standard conditions only if

$$\frac{[\text{lactate}]\,[\text{NAD}^+]}{[\text{pyruvate}]\,[\text{NADH}]} < 160^2 = \textbf{25,600.}$$

(e) From the forgoing,

$$\frac{[\text{lactate}]}{[\text{pyruvate}]} < \frac{25,600}{\left(\dfrac{[\text{NAD}^+]}{[\text{NADH}]}\right)} < \frac{25,600}{1000}$$

so that

$$\frac{[\text{lactate}]}{[\text{pyruvate}]} < \textbf{25.6}$$

9. (a) In analogy with Figure 16-25, TPP would catalyze the decarboxylation of the carboxyl group α to the carbonyl group but not that β to the carbonyl group. Rather, this reaction proceeds by enolization of the keto function.

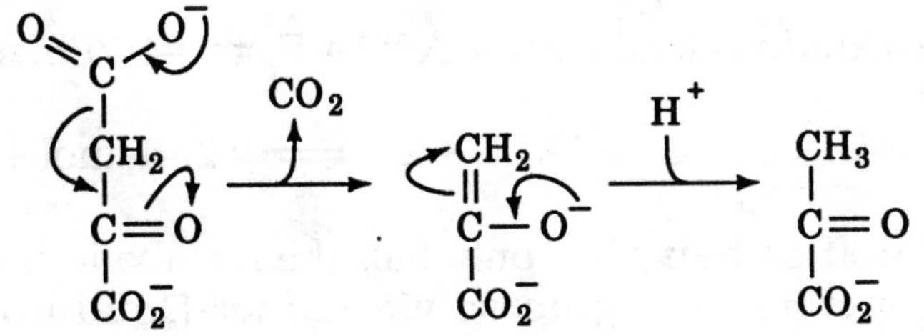

(The reaction would be facilitated by Schiff base formation with the carbonyl group.)

(b) This reaction involves the decarboxylation of an α-ketoacid.

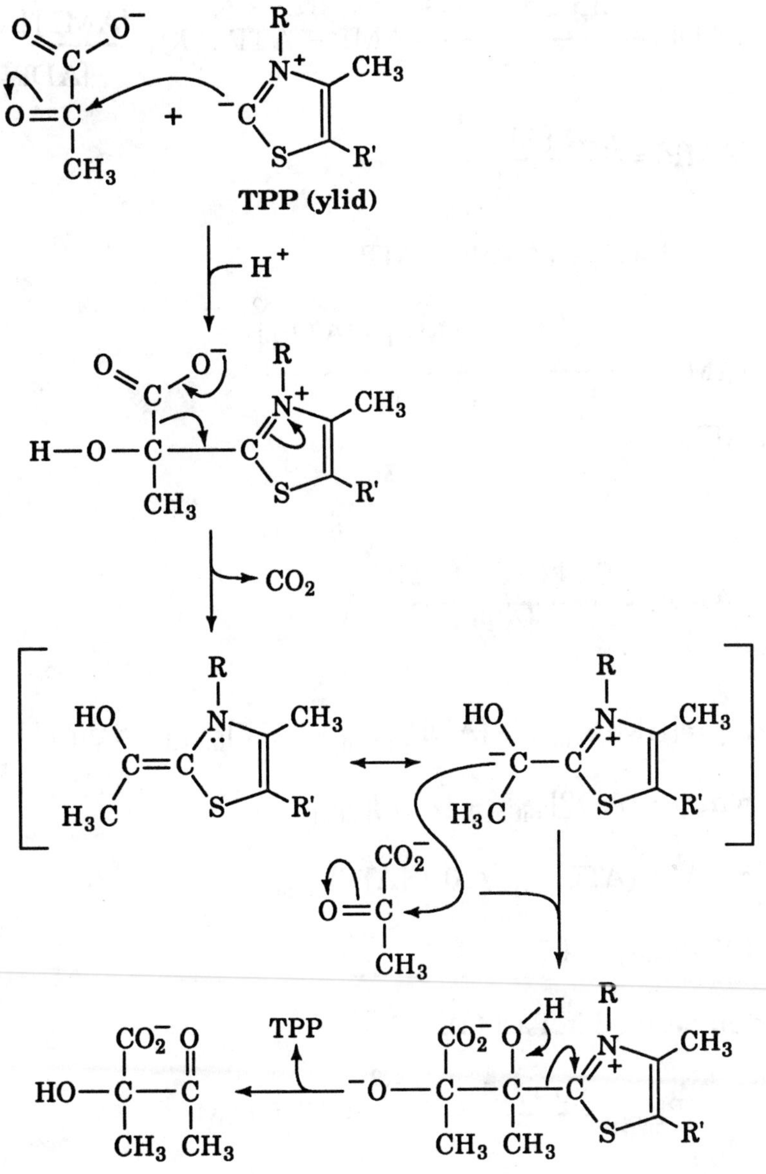

Thus, this reaction can be catalyzed by TPP.

10. (a)

$$\text{Entner-Doudoroff: Glucose} + \text{ADP} + P_i \rightleftharpoons 2\text{Ethanol} + 2CO_2 + \text{ATP}$$

$$\text{Glycolysis: Glucose} + 2\text{ADP} + 2P_i \rightleftharpoons 2\text{Ethanol} + 2CO_2 + 2\text{ATP}$$

(b)   The Entner-Doudoroff pathway has only half the metabolic efficiency of glycolysis; that is, one ATP is generated per glucose via Entner-Doudoroff *vs* two ATP's for glycolysis. Hence, organisms with the Entner-Doudoroff pathway would compete poorly with organisms having glycolysis, at least under anaerobic conditions.

11.

$$2\text{ADP} \xrightarrow{\substack{\text{adenylate} \\ \text{kinase}}} \text{AMP} + \text{ATP} \qquad K = \frac{[\text{AMP}][\text{ATP}]}{[\text{ADP}]^2} = 0.44$$

$$[\text{AMP}] = \frac{0.44\,[\text{ADP}]^2}{[\text{ATP}]}$$

$$[\text{ADP}] = A_T - [\text{AMP}] - [\text{ATP}]$$

$$[\text{AMP}] = \frac{0.44\left(A_T - [\text{AMP}] - [\text{ATP}]\right)^2}{[\text{ATP}]}$$

But $[\text{ATP}] \gg [\text{AMP}]$.

Therefore,

$$[\text{AMP}] = \frac{0.44\,(A_T - [\text{ATP}])^2}{[\text{ATP}]}$$

(b)

$$A_T = [\text{AMP}]_{\text{initial}} + [\text{ADP}]_{\text{initial}} + [\text{ATP}]_{\text{initial}} = 5\,mM$$

$$5\,mM \approx [\text{ADP}]_{\text{initial}} + [\text{ATP}]_{\text{initial}}$$

$$5\,mM = [\text{ATP}]_{\text{initial}}/10 + [\text{ATP}]_{\text{initial}}$$

$$[\text{ATP}]_{\text{initial}} = 4.55\,mM$$

Using the equation derived in Part (a),

$$[\text{AMP}]_{\text{initial}} = \frac{0.44\,(5 - 4.55)^2}{4.55} = 0.020\,mM$$

A 10% decrease in [ATP] must be compensated by an increase in [ADP] + [AMP] since A_T is constant. Assume most of the change goes into [ADP] so that $[\text{ATP}] \gg [\text{AMP}]$. Then,

$$[ATP]_{new} = 0.9\,[ATP]_{initial} = 0.9 \times 4.55 \text{ mM} = 4.09 \text{ mM}$$

$$[AMP]_{new} = \frac{0.44\,(5 - 4.09)^2}{4.09} = 0.089 \text{ m}M$$

$$\frac{[AMP]_{new}}{[AMP]_{initial}} = \frac{0.089}{0.020} = 4.45 = 445\%$$

12. Muscle rigidity and heat generation require energy metabolism. The end product of glycolysis in mammals is lactic acid. Also, the overall hydrolysis of ATP, once it has run out, yields some acid:

$$ATP + H_2O \rightleftharpoons ADP + HPO_4^{2-} + H^+$$

(although there is never much ATP present). The meat, as a result, is acidic.

Chapter 17
GLYCOGEN METABOLISM

1. A reducing end of a polysaccharide chain has a free C(1)—OH group. Hence, a glycogen molecule has but one reducing end, no matter what its degree of polymerization or branching.

2. Glycogen synthesis:

$$\text{Glucose} + \text{ATP} \xrightarrow{\text{hexokinase}} \text{G6P} + \text{ADP}$$

$$\text{G6P} \xrightarrow{\text{phosphoglucomutase}} \text{G1P}$$

$$\text{G1P} + \text{UTP} \xrightarrow{\text{UDP-glucose pyrophosphorylase}} \text{UDPG} + \text{PP}_i$$

$$\text{PP}_i + \text{H}_2\text{O} \xrightarrow{\text{inorganic pyrophosphatase}} 2\text{P}_i$$

$$\text{UDPG} + \text{Glycogen}(n-1 \text{ residues}) \xrightarrow[\text{synthase}]{\text{glycogen}} \text{UDP} + \text{Glycogen}(n \text{ residues})$$

$$\text{UDP} + \text{ATP} \xrightarrow{\text{nucleoside diphosphate kinase}} \text{UTP} + \text{ADP}$$

Sum: $\text{Glucose} + 2\text{ATP} + \text{Glycogen}(n-1 \text{ residues}) + \text{H}_2\text{O}$

$$\longrightarrow \text{Glycogen}(n \text{ residues}) + 2\text{ADP} + 2\text{P}_i$$

Branching requires no energy (ATP) input.

Glycogen breakdown:

$$\text{Glycogen}(n \text{ residues}) + 0.9\,P_i + 0.1\,H_2O \xrightarrow{\text{glycogen phosphorylase + debranching enzyme}}$$

$$\text{Glycogen}(n-1 \text{ residues}) + 0.9\,\text{G1P} + 0.1\,\text{Glucose}$$

To enter glycolysis, the latter products react as follows:

$$0.9\,\text{G1P} \xrightarrow{\text{phosphoglucomutase}} 0.9\,\text{G6P}$$

$$0.1\,\text{Glucose} + 0.1\,\text{ATP} \xrightarrow{\text{hexokinase}} 0.1\,\text{G6P} + 0.1\,\text{ADP}$$

Summing up all these reactions:

$$\text{Glucose} + 2.1\,\text{ATP} + 1.1\,H_2O \longrightarrow \text{G6P} + 2.1\,\text{ADP}$$

The direct metabolism of glucose, to reach the same stage, requires the reaction.

$$\text{Glucose} + \text{ATP} \xrightarrow{\text{hexokinase}} \text{G6P} + \text{ADP}$$

Hence, the synthesis and breakdown of glycogen consumes (2.1 – 1) = 1.1 ATP per glucose unit. The fractional energetic cost of glycogen synthesis is therefore 1.1/38 = **0.029 of the energy of direct glucose metabolism.**

3. (a) Ca^{2+} binds to the calmodulin component of phosphorylase kinase so as to activate this enzyme to phosphorylate glycogen phosphorylase thereby **increasing the rate of degradation.** Similarly, phosphorylase kinase phosphorylates glycogen synthase, thereby **decreasing the rate of glycogen synthesis.**

(b) ATP allosterically inhibits glycogen phosphorylase so that its presence **decreases glycogen breakdown.** *m*-Glycogen synthase *b* is already inactive with physiological amounts of ATP and thus is unaffected by more. [ATP] increases therefore have **no effect on glycogen synthesis.**

(c) Adenylate cyclase synthesizes the cAMP that stimulates the phosphorylation of phosphorylase kinase (activates glycogen degradation and deactivates its synthesis), glycogen synthase (deactivates glycogen synthesis), and phosphoprotein phosphatase inhibitor (prevents deactivation of glycogen degradation and activation of its synthesis). The resulting cAMP shortage therefore **prevents glycogen breakdown and stimulates its synthesis.**

(d) Epinephrine stimulates adenylate kinase in muscle and liver. It therefore has the opposite effects of (c): **it stimulates glycogen breakdown and inhibits its synthesis.**

(e) Increasing [AMP] stimulates glycogen phosphorylase thereby **increasing the rate of**

glycogen breakdown. AMP has no direct effect on glycogen synthase and its increase in [ATP] through the adenylate kinase reaction likewise has **no effect on the synthesis of glycogen.**

4.

$$v_o = \frac{V_{max}[S]}{K_M + [S]}$$

When $K_M \ll [S]$, as it is with hexokinase ($K_M < 0.1$ mM vs $[S] = 5$ mM; see Fig. 17-5), $v_o \approx V_{max}$ so that this quantity is insensitive to $[S]$, that is, the enzyme is saturated. However, when $K_M \approx [S]$, as it is with glucokinase ($K_M \approx 10$ mM; *note that this quantity was erroneously stated to be 5 mM in the first printing of the text*), then both of these quantities are significant in the Michaelis-Menten equation; that is, v_o varies with $[S]$.

$$r = \frac{v_o(\text{glucokinase})}{v_o(\text{hexokinase})} = \frac{V_{max}(\text{glucokinase})\,(0.1 + [\text{glucose}])}{V_{max}(\text{hexokinase})\,(10^* + [\text{glucose}])}$$

*5 in the first printing.

But according to our assumption,

$$V_{max}(\text{glucokinase}) = V_{max}(\text{hexokinase})$$

so that,

$$r = \frac{0.1 + [\text{glucose}]}{10^* + [\text{glucose}]}$$

At 2 mM glucose: $\qquad r = \frac{0.1 + 2}{10^* + 2} = 0.17$ (**0.30 in the first printing**).

At 5 mM glucose: $\qquad r = \frac{0.1 + 5}{10^* + 5} = 0.34$ (**0.51 in the first printing**).

At 25 mM glucose: $\qquad r = \frac{0.1 + 25}{10^* + 25} = 0.72$ (**0.84 in the first printing**).

*5 in the first printing

5. From equation [17.A3],

$$A = 1 + \frac{3k_r \times 3[R]_T \times 3K_f(3e_2 + e_2)}{K_f k_r [R]_T e_2}$$

$$A = 1 + (3 \times 3 \times 3 \times 4) = 109$$

6. Bicylic cascades have properties resembling monocyclic cascades but, in comparison, have even greater amplification potential, can respond to a greater number of allosteric stimuli, and can exhibit greater flexibility in their control patterns.

7. Muscle uses glycogen breakdown for the rapid acquisition of metabolic energy. Since muscle only stores a few seconds worth of ATP and creatine (an ATP buffer; Section 15-4C), glycogen must be rapidly mobilized when there is a need for it. Liver, on the other hand, functions to maintain a steady level of blood glucose, a quantity that fluctuates over minutes and hours rather than seconds. Muscle glycogen phosphorylase is therefore better adapted to its function if it responds more quickly to external stimuli.

8. Intramolecular autophosphorylation would probably serve to maintain a certain basal level of phosphorylase kinase activity in the presence of active phosphoprotein phosphatase. Intermolecular autophosphorylation probably increases the amplification factor on the cAMP stimulation of glycogen breakdown and inhibition of glycogen synthesis. This is because autophosphorylation provides the same effect as phosphorylation by cAMP-dependent protein kinase thereby enhancing the effect of this enzyme.

9. von Gierke's disease, a deficiency in glucose-6-phosphatase, leads to a build-up of G6P in the liver (this substance cannot pass through liver cell membranes). The abnormally high [G6P] allosterically inhibits glycogen phosphorylase but stimulates glycogen synthase leading to massive glycogen accumulation, thereby accounting for liver enlargement. Liver glycogen normally serves as a blood glucose buffer after meals have been digested. Since the liver cannot convert G6P to glucose, von Gierkes disease leads to fasting hypoglycemia.

10. The deficiency is in branching enzyme (Type IV, Andersen's disease). The glycogen has very few branches and hence very long $\alpha(1 \rightarrow 4)$-linked branches. (Normal glycogen has branches every ~10 residues, yielding a G1P/glucose ratio of ~10 on glycogenolysis.)

Chapter 18
TRANSPORT THROUGH MEMBRANES

1.

$$\Delta \overline{G} = RT \ln \frac{[G_{in}]}{[G_{out}]} \quad \text{for transport into the cell.}$$

$$T = 27°C = 310 \text{ K}$$

$$\Delta \overline{G} = 8.314 \times 10^{-3} \text{ kJ} \cdot \text{mol}^{-1} \text{K}^{-1} \times 310 \text{ K} \times \ln \frac{10^{-4} M}{10^{-2} M}$$

$$\Delta \overline{G} = -11.9 \text{ kJ} \cdot \text{mol}^{-1}$$

2.

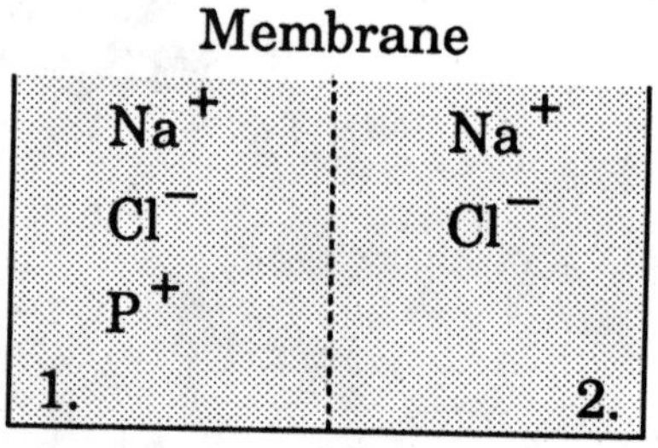

Membrane

Na$^+$	Na$^+$
Cl$^-$	Cl$^-$
P$^+$	
1.	2.

Let us call the compartment containing the protein P$^+$ side 1 and the compartment lacking P$^+$ side 2. We begin by calculating the concentrations of the various species on the two sides of the membrane.

For Na$^+$:

$$\Delta G_{Na} = G_{Na}(2) - G_{Na}(1) = (G^{\circ}_{Na} + RT \ln [Na^+(2)]) - (G^{\circ}_{Na} + RT \ln [Na^+(1)])$$

$$\Delta G_{Na} = RT \ln \frac{[Na^+(2)]}{[Na^+(1)]}$$

For Cl$^-$, similarly:

$$\Delta G_{Cl} = RT \ln \frac{[Cl^-(2)]}{[Cl^-(1)]}$$

At equilibrium,

$$\Delta G_{Na} + \Delta G_{Cl} = 0$$

so that

$$RT \ln \frac{[Na^+(2)] \, [Cl^-(2)]}{[Na^+(1)] \, [Cl^-(1)]} = 0$$

Thus,

$$[Na^+(2)][Cl^-(2)] = [Na^+(1)][Cl^-(1)] \tag{1}$$

Electroneutrality demands

$$[P^+] + [Na^+(1)] = [Cl^-(1)] \tag{2}$$

and

$$[Na^+(2)] = [Cl^-(2)] = c \tag{3}$$

Substituting (3) into (1):

$$[Cl^-(1)] = \frac{c^2}{[Na^+(1)]} \tag{4}$$

which, when substituted into (2), yields

$$\frac{[P^+] \, [Na^+(1)] + [Na^+(1)]^2 - c^2}{[Na^+(1)]} = 0$$

or

$$[Na^+(1)]^2 + [P^+] \, [Na^+(1)] - c^2 = 0$$

Solving for [Na$^+$(1)] using the quadratic formula,

$$[Na^+(1)] = \frac{-[P^+] + \sqrt{[P^+]^2 + 4c^2}}{2} \tag{5}$$

(Here we have chosen only the positive root since the negative root yields the physically absurd result of a negative concentration.)

Conservation of mass indicates that

$$\{[Na^+(1)] + c\}V = 0.1V \qquad \text{where } V \text{ is the compartment volume.}$$

so that

$$[Na^+(1)] = 0.1M - c \tag{6}$$

Likewise,

$$[P^+] = 0.1M \tag{7}$$

Substituting (6) and (7) into (5),

$$0.1 - c = \frac{-0.1 + \sqrt{0.1^2 + 4c^2}}{2}$$

$$\sqrt{0.01 + 4c^2} = 0.2 - 2c + 0.1 = 0.3 - 2c$$

$$0.01 + 4c^2 = 0.09 - 1.2c + 4c^2$$

$$1.2c = 0.09 - 0.01 = 0.08$$

$$c = 0.08/1.2 = \mathbf{0.0667M} = [Na^+(2)] = [Cl^-(2)]$$

From equation (6):

$$[Na^+(1)] = 0.1 - c = 0.1 - 0.0667 = \mathbf{0.0333M}$$

From equation (2):

$$[Cl^-(1)] = [Na^+(1)] + [P^+] = 0.0333 + 0.1 = 0.1333M$$

The membrane potential is calculated from equation [18.3].

$$\Delta \overline{G}_{Na} = RT\ln\frac{[Na^+(1)]}{[Na^+(2)]} + Z_{Na}F\Delta\Psi = 0 \qquad \text{at equilibrium}$$

$$\Delta\Psi = \frac{-RT}{F}\ln\frac{[Na^+(1)]}{[Na^+(2)]} = \frac{-8.314 \; J\bullet mol^{-1}K^{-1}\,(273+25)}{96,494 \; J\bullet V^{-1}mol^{-1}}\ln\left(\frac{0.0333M}{0.0667M}\right)$$

$\Delta\Psi = +\,\mathbf{0.018}$ **V** (positive in compartment 1).

3. Gramicidin A transports 10^7 Na^+ ions$\bullet s^{-1}$ channel^{-1}

$$V = 80 \; \mu m^3 = 80 \times (10^{-4} \; cm)^3 = 8.0 \times 10^{-11} \; mL$$

$$10 \; mM = 10^{-5} \; mol\bullet mL^{-1} = 10^{-5} \; mol\bullet mL^{-1} \times 6.02 \times 10^{23} \; ions\bullet mol^{-1}$$

$$= 6.02 + 10^{18} \; ions\bullet mL^{-1}$$

Number of ions to be transferred $= 8.0 \times 10^{-11} \; mL \times 6.02 \times 10^{18} \; ions\bullet mL^{-1}$

$$= 4.8 \times 10^8 \; ions\bullet mL^{-1}$$

$$\text{Time} = \frac{4.8 \times 10^8 \; ions}{10^7 \; ions\bullet s^{-1}} = \mathbf{48 \; s}$$

4. ATP hydrolysis occurs via a pentacovalent phosphorus atom with trigonal bipyramidal coordination. Vanadate ion, which has identical coordination geometry, resembles the transition state and binds at the enzyme's active site thereby inhibiting ATP hydrolysis. Vanadate is an example of a transition state analog.

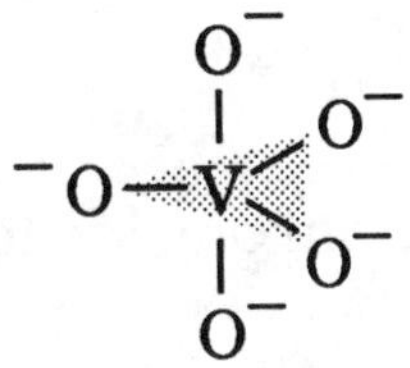

Trigonal bipyramid

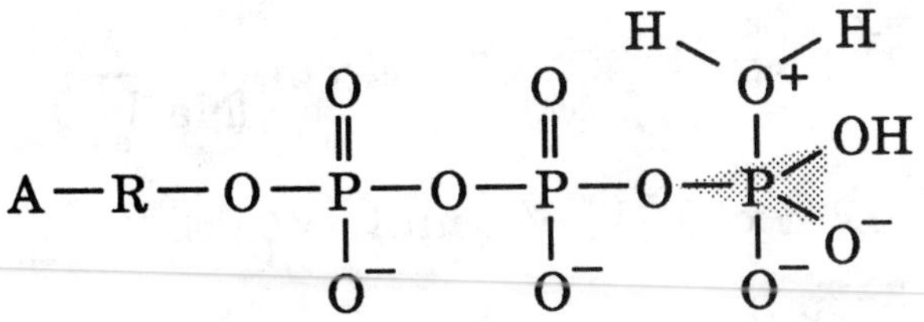

Trigonal bipyramid
transition state
for ATP hydrolysis

5.

$$H^+_{in} \rightleftharpoons H^+_{out}$$

$$\Delta G = RT \ln \frac{[H^+]_{out}}{[H^+]_{in}} + Z_H F \Delta\Psi$$

$$\Delta\Psi = \Psi_{out} - \Psi_{in} = +0.06\ V = 0.06\ J \bullet C^{-1}$$

$$Z_H = 1$$

Assuming the physiological temperature of 37°C = 310K,

$$\Delta G = 8.314 \times 10^{-3}\ kJ \bullet mol^{-1}\ K^{-1} \times 310\ K \times \ln\frac{0.18M}{10^{-7}M} +$$

$$1 \times 96{,}494\ C \bullet mol^{-1} \times 0.06 \times 10^{-3}\ kJ \bullet C^{-1} = 37.1 + 5.79 = 42.9 kJ \bullet mol^{-1}$$

Thus, since ΔG for ATP hydrolysis is –31.5 kJ•mol^{-1}, and 42.9/31.5 = 1.4, **at least 2 mol of ATP must be hydrolyzed per mole of H$^+$ transported.**

6. 100 mV potential difference across a 100Å membrane corresponds to

$$\frac{0.1\ V}{100\ \mathring{A} \times 10^{-10}\ m \bullet \mathring{A}^{-1} \times 100\ cm \bullet m^{-1}} = 100{,}000\ V \bullet cm^{-1}$$

This is an enormous potential difference on the macroscopic scale. For instance, electrical transmission lines carrying 100,000 V are supported by insulators more than 1 m long to prevent them from arcing to their support towers.

7.

$$\Delta G = RT\ln\frac{[Na^+]_{in}}{[Na^+]_{out}} + nZ_{Na}F\Delta\Psi$$

where $\Delta\Psi = \Psi(in) - \Psi(out)$ for migration from outside to inside the membrane.

Assuming $T = 37°C = 310\ K$,

$$-1.9\ kJ \bullet mol^{-1} = 8.314 \times 10^{-3}\ kJ \bullet mol^{-1}\ K^{-1} \times 310\ K \times \ln\frac{[Na^+]_{in}}{[Na^+]_{out}}$$

$$+\ 1 \times 1 \times 96{,}494 \times 10^{-3}\ kJ \bullet V^{-1}\ mol^{-1} \times 0.060\ V$$

$$-1.9 = 2.58\ \ln\frac{[Na^+]_{in}}{[Na^+]_{out}} + 5.8$$

$$\frac{-1.9 - 5.8}{2.58} = \ln\frac{[Na^+]_{in}}{[Na^+]_{out}}$$

$$-2.98 = \ln\frac{[Na^+]_{in}}{[Na^+]_{out}}$$

$$[Na^+]_{out} = 260 \text{ m}M$$

$$[Na^+]_{in} = 0.051 \, [Na^+]_{out} = 0.051 \times 260 \text{ m}M = \textbf{13.3 m}\textit{M}$$

8. In nonmediated diffusion, the initial rate of uptake varies linearly with the external concentration. Mediated diffusion obeys saturation kinetics so that at high external concentrations the rate no longer increases with increasing external concentration.
   Hence, as can be seen by inspection of the table, leucine enters the cell via mediated diffusion whereas ethylene glycol enters it via nonmediated diffusion.

9. (a) Ethanol: CH_3CH_2OH

   A neutral lipophilic molecule that therefore does not require mediation to cross a membrane.

   (b) Glycine $\overset{+}{H_3N}-CH_2-COO^-$

   A zwitterionic molecule with low solubility in nonpolar media that therefore requires mediation to cross a membrane.

   (c) Cholesterol

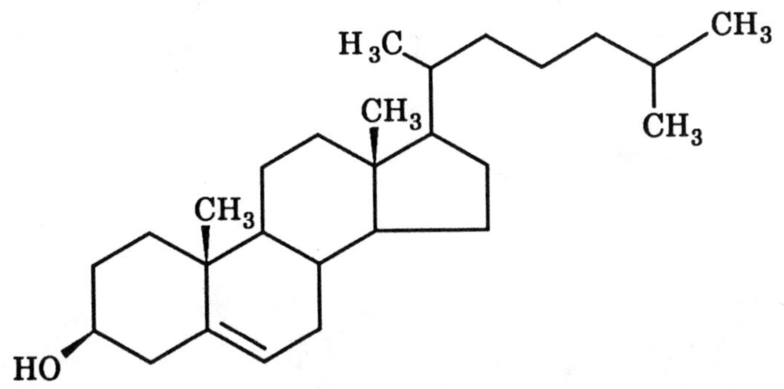

   A neutral amphiphilic molecule that must therefore flip-flop in order to traverse a lipid bilayer. This process is apparently catalyzed by membrane proteins (Section 11-3F.)

   (d) ATP

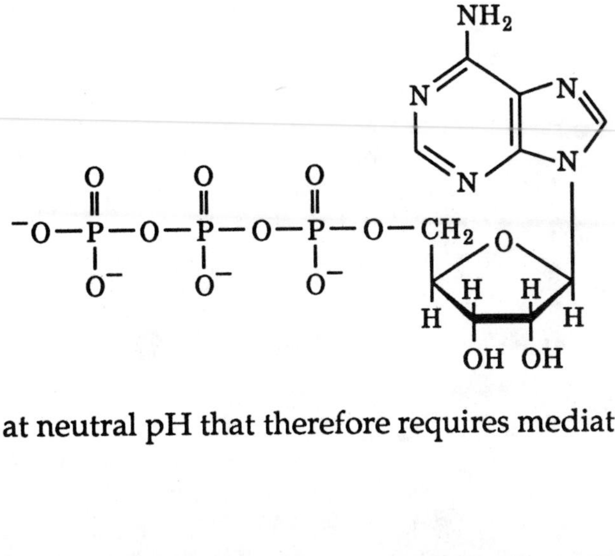

   An anionic species at neutral pH that therefore requires mediation to cross a membrane.

10.

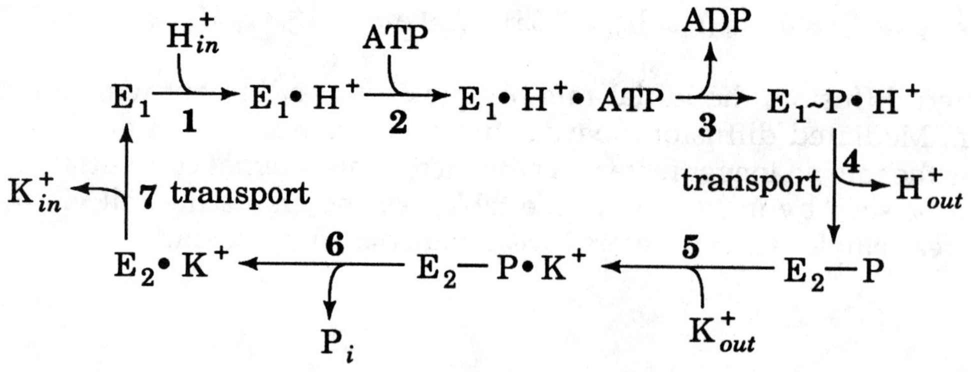

$$
\begin{array}{ccccccc}
& H_{in}^{+} & & ATP & & ADP & \\
E_1 & \xrightarrow{\quad 1 \quad} & E_1 \cdot H^{+} & \xrightarrow{\quad 2 \quad} & E_1 \cdot H^{+} \cdot ATP & \xrightarrow{\quad 3 \quad} & E_1 \sim P \cdot H^{+} \\
\uparrow & & & & & & \downarrow\ 4 \\
K_{in}^{+} \leftarrow 7\ \text{transport} & & & & \text{transport} & & H_{out}^{+} \\
E_2 \cdot K^{+} & \xleftarrow{\quad 6 \quad} & E_2 - P \cdot K^{+} & \xleftarrow{\quad 5 \quad} & E_2 - P & & \\
& \downarrow & & & & & \\
& P_i & & K_{out}^{+} & & &
\end{array}
$$

The coupling of ATP hydrolysis with H^+ transport requires that ATP hydrolysis only occur as the H^+ is transported "uphill" and that "downhill" H^+ transport only occurs concomitantly with ATP synthesis. This is accomplished by an ordered sequential mechanism in which ATP binds to E_1 only after H^+ and E_2—P hydrolyzes, which can occur only after K^+ binds. Mutual destabilization occurs at steps 3+4 and at steps 5+6.

Chapter 19
CITRIC ACID CYCLE

---

1. Glucose labeled with ^{14}C at C(2) is converted to acetyl-CoA which is ^{14}C-labeled at its carbonyl group. As shown in Figure 19-1, this carbon atom, identified by (‡), becomes C(5) of α-ketoglutarate and is scrambled between C(1) and C(4) of succinate. This atom will not be released as $^{14}CO_2$ during the first round of isocitrate dehydrogenase and α-ketoglutarate dehydrogenase reactions. Two turns of the citric acid cycle are therefore necessary for complete conversion of radioactivity to $^{14}CO_2$.

Pyruvate, ^{14}C-labeled at its methyl group, becomes methyl-labeled acetyl-CoA. The labeled carbon atom becomes C(4) of α-ketoglutarate and is scrambled between C(2) and C(3) of succinate. This atom will not be released during the first turn of the cycle as $^{14}CO_2$. Neither will it be released during the second turn of the cycle since, as we have seen above, C(1) and C(4) of oxaloacetate are the two atoms released as CO_2 during the cycle. However, C(2) and C(3) of oxaloacetate become C(2) and C(3) of α-ketoglutarate which then are scrambled throughout the four carbon atoms of succinate and the ensuing oxaloacetate. Thus, in the third turn of the cycle, half the radioactivity (at C(1) and C(4) of oxaloacetate) will be released as $^{14}CO_2$. In each subsequent turn, half the remaining radioactivity will be released.

Note that these problems demonstrate that although acetyl groups are formally oxidized by the citric acid cycle, it is actually the carbon atoms of oxaloacetate that become CO_2. The carbon atoms of an acetyl group that are incorporated into oxaloacetate during a given turn of the cycle are not oxidized until subsequent turns of the cycle.

2. Isocitrate + NAD^+ $\rightleftharpoons$ α-ketoglutarate + NADH + CO_2 + H^+

$[H^+]$ and $[CO_2]$ are considered to be 1.

$$\Delta G = \Delta G^{\circ\prime} + RT\ln\frac{[NADH][\alpha\text{-ketoglutarate}]}{[NAD^+][\text{isocitrate}]}$$

$$= -21 + 8.31 \times 10^{-3} \times 298 \times \ln\frac{0.1 \times 10^{-3}}{8 \times 0.02 \times 10^{-3}}$$

$$= -21 + 2.48 \ln 0.63 = -22.17 \text{ kJ} \cdot \text{mol}^{-1}$$

With such a large negative free energy of reaction under physiological conditions, isocitrate dehydrogenase is an excellent candidate for metabolic control.

3. (1) Isocitrate dehydrogenase NAD^+

   (2) α-Ketoglutarate dehydrogenase NAD^+

   (3) Succinate dehydrogenase   FAD

   (4) Malate dehydrogenase NAD^+

   (1)

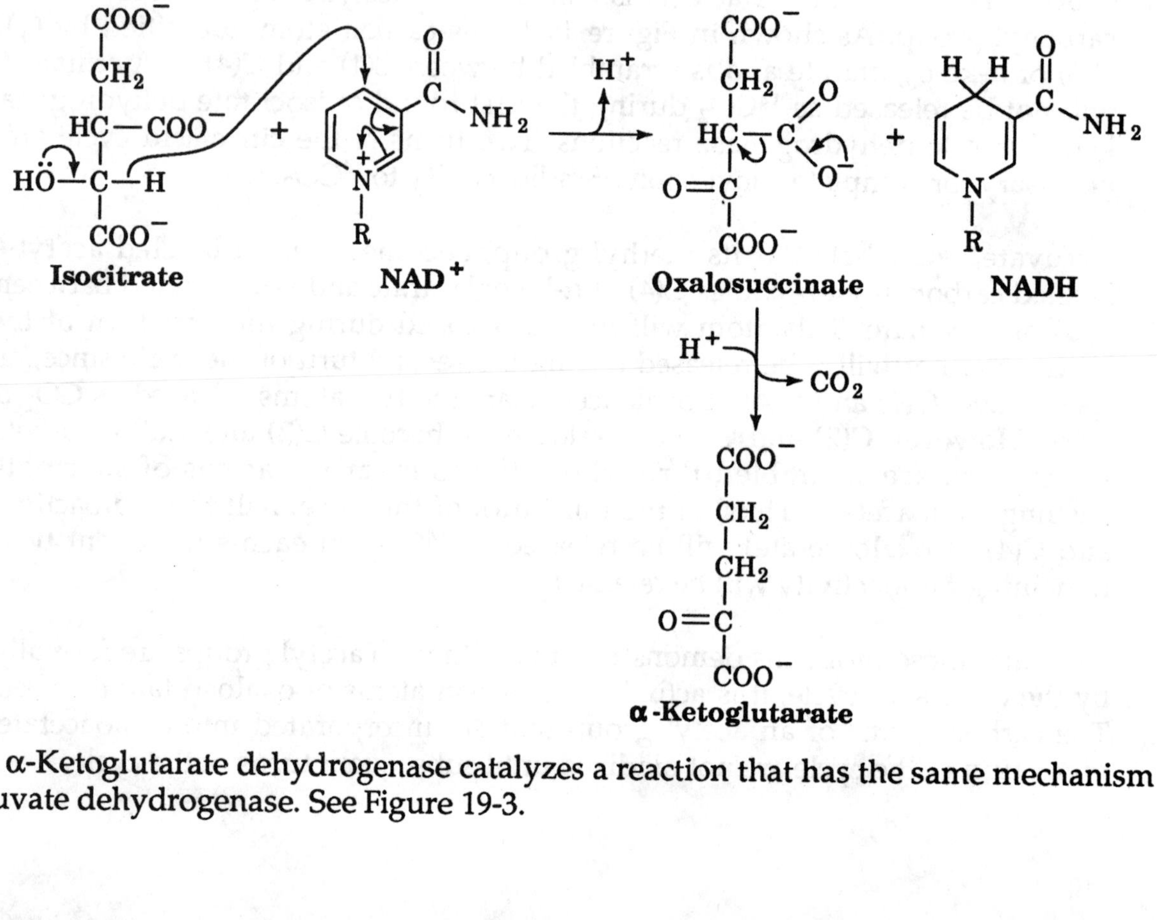

(2) α-Ketoglutarate dehydrogenase catalyzes a reaction that has the same mechanism as pyruvate dehydrogenase. See Figure 19-3.

(3)

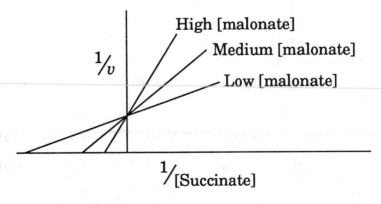

Succinate   **FAD**

(4)   The malate dehydrogenase reaction mechanism is identical to that of the first step of the isocitrate dehydrogenase reaction. See (1) above.

4. One proposal that would account for the fact that an enzyme-bound carbanion does not exchange ^3H with ^3H_2O is that the enzyme only forms the carbanion in the presence of the second substrate, oxaloacetate, which changes the enzyme's conformation so as to facilitate carbanion formation. [Credence is given to this explanation by the observation that addition of (S)-malate [but not (R)-malate] is known to cause measurable exchange of ^3H from ^3H_2O into acetyl-CoA in the presence of citrate synthase. (S)-malate resembles oxaloacetate enough to stimulate a conformational change on the enzyme facilitating acetyl-CoA carbanion formation.]

5.

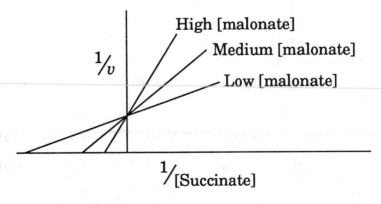

6. Assuming a sufficient supply of acetyl-CoA, raising the oxaloacetate concentration has the effect of raising the concentration of all of the citric acid cycle intermediates,

including succinate. Since malonate is a competitive inhibitor of succinate dehydrogenase, increasing the succinate concentration can relieve this inhibition.

7. Only (d), acetyl-CoA, undergoes net oxidation by the citric acid cycle. All the others are regenerated by the first turn of the cycle.

8.

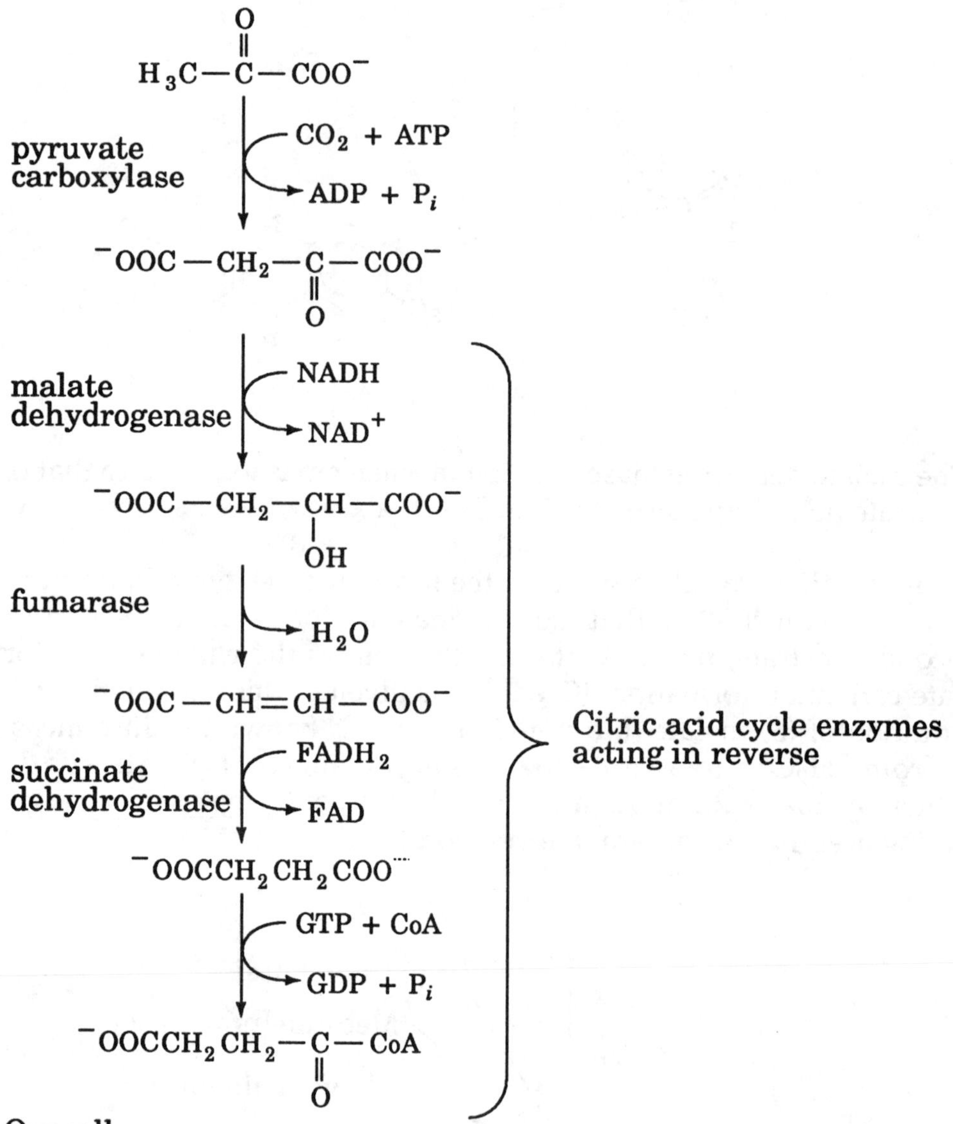

$$H_3C-\overset{\overset{\displaystyle O}{\|}}{C}-COO^-$$

pyruvate carboxylase

$CO_2 + ATP$
$ADP + P_i$

$$^-OOC-CH_2-\overset{\overset{\displaystyle}{}}{C}-COO^-$$
$$\underset{O}{\|}$$

malate dehydrogenase

$NADH$
NAD^+

$$^-OOC-CH_2-\underset{\underset{\displaystyle OH}{|}}{CH}-COO^-$$

fumarase

H_2O

$$^-OOC-CH=CH-COO^-$$

succinate dehydrogenase

$FADH_2$
FAD

$$^-OOCCH_2CH_2COO^-$$

$GTP + CoA$
$GDP + P_i$

$$^-OOCCH_2CH_2-\overset{\overset{\displaystyle}{}}{C}-CoA$$
$$\underset{O}{\|}$$

Citric acid cycle enzymes acting in reverse

Overall:

Pyruvate + ATP + CO$_2$ + NADH + FADH$_2$ + GTP + CoA

$\longrightarrow$ Succinyl-CoA + ADP + NAD$^+$ + FAD + GDP + 2P$_i$

9. While succinyl-CoA can be synthesized from pyruvate by the reversal of the citric acid cycle enzymes shown in the solution to Problem 8, the α-ketoglutarate dehydrogenase

reaction is irreversible, preventing a continuation of that pathway to produce α-ketoglutarate. Instead, the forward direction of the citric acid cycle must be utilized.

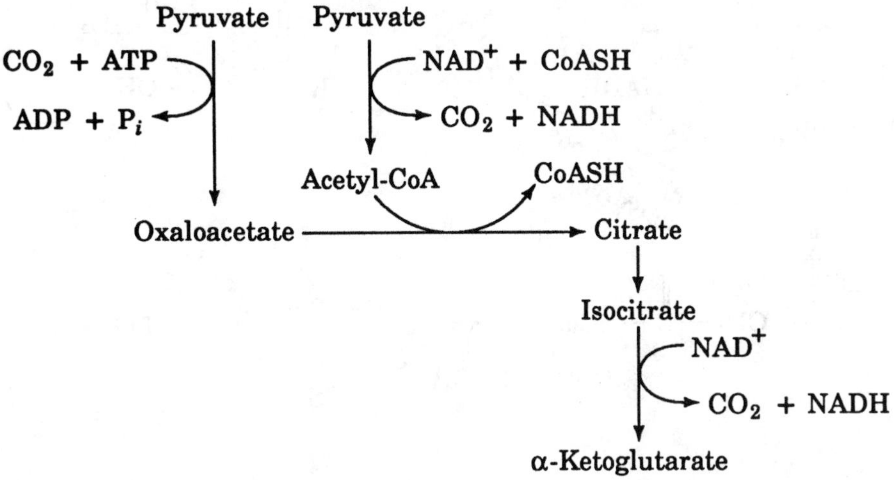

Overall:

$$2\text{Pyruvate} + \text{ATP} + 2\text{NAD}^+ \longrightarrow \alpha\text{-Ketoglutarate} + 2\text{NADH} + CO_2 + \text{ADP} + P_i$$

10. (a) Lipoic acid is bound to enzymes through an amide bond with the ε-amino group of a Lys residue. It is often referred to as lipoamide for this reason.

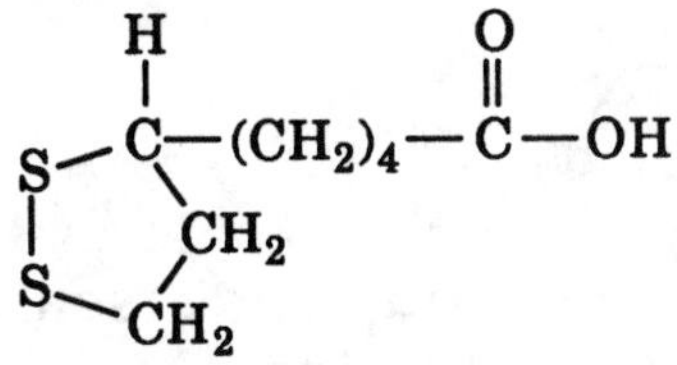

Lipoic acid

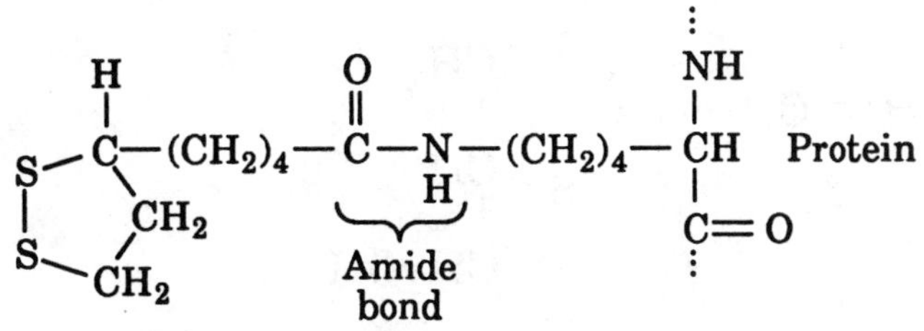

Lipoic acid–protein complex

(b) Lipoic acid participates in the oxidative decarboxylation of α-ketoacids such as pyruvate as part of a multi-enzyme complex. It accepts a decarboxylated carbanion of the substrate from thiamine pyrophosphate (TPP). The process involves oxidation of the substrate to a thioester and reduction of the disulfide of lipoic acid. The acyl group is then transferred to the thiol of coenzyme A, leaving the dihydrolipoyl group which must be reoxidized before the catalytic cycle can be repeated.

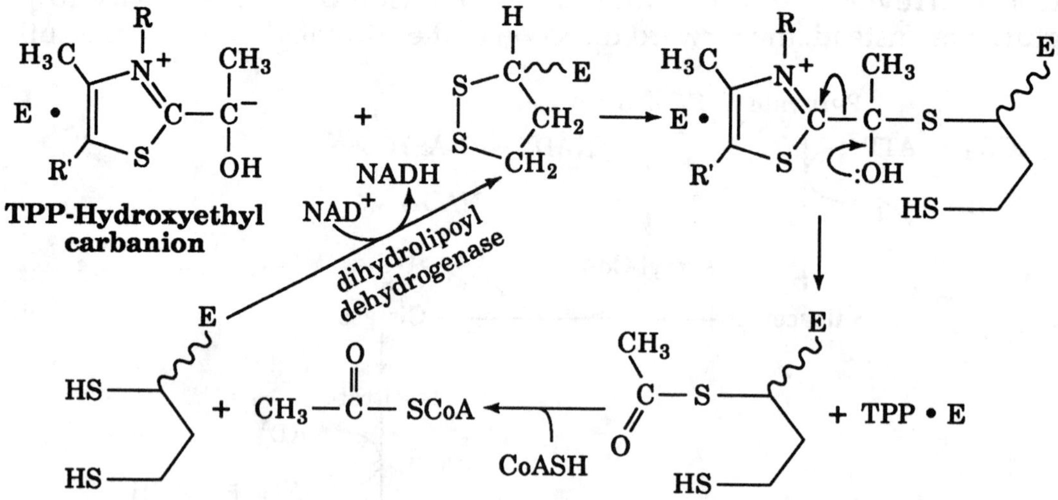

11. Arsenic poisoning by lewisite is thought to be caused by the formation of a complex between the arsenic-containing compound and the dithiol of the dihydrolipoyl group in the pyruvate dehydrogenase multienzyme complex.

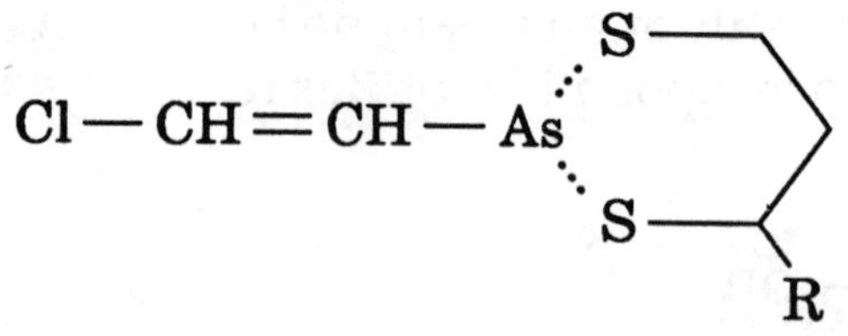

BAL would compete with the dihydrolipoyl group for arsenic compounds thus protecting lipoic acid-containing enzymes from inactivation.

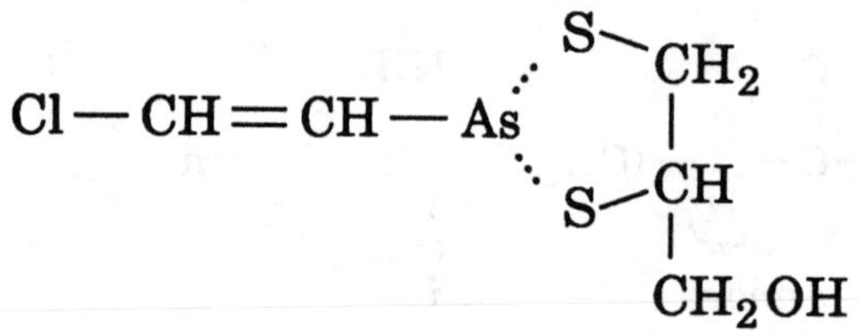

Chapter 20

ELECTRON TRANSPORT AND OXIDATIVE PHOSPHORYLATION

1. Cytochrome a > cytochrome c > CoQ > FAD > NAD$^+$

2. The oxidation of succinate to fumarate catalyzed by succinate dehydrogenase involves the reduction of FAD to FADH$_2$. Electrons from FADH$_2$ enter the electron transport chain at CoQ bypassing the first of the three energy conserving steps in oxidative phosphorylation. The oxidation of malate to oxaloacetate catalyzed by malate dehydrogenase involves reduction of NAD$^+$ to NADH. Electrons from NADH enter the electron transport chain at NADH dehydrogenase and pass through all three energy conserving steps in oxidative phosphorylation. Each energy conserving step supplies the free energy necessary for the synthesis of ATP.

3. $$1/2 O_2 + FADH_2 \rightleftharpoons FAD + H_2O$$

From Table 15-4,

$$\Delta E^{\circ\prime} = +0.815 \text{ V} - (-0.219 \text{ V}) = 1.034 \text{ V}$$

$$\Delta G^{\circ\prime} = -nF\Delta E^{\circ\prime} = -2 \times 96.5 \times 1.034 = -199.6 \text{ kJ} \bullet \text{mol}^{-1}$$

$$2P_i + 2ADP \rightleftharpoons 2ATP + 2H_2O \quad \Delta G^{\circ\prime} = 2 \times 30.5 = +61.0 \text{ kJ} \bullet \text{mol}^{-1}$$

$$\text{Efficiency} = \frac{61.0}{199.6} = 0.31 \text{ or } 31\%$$

31% of the free energy released on oxidizing $FADH_2$ by O_2 is utilized in the synthesis of 2 moles of ATP. The thermodynamic efficiency of this process is therefore **31%**.

4. The most toxic effect of CN^- is its binding to the cytochromes of cytochrome oxidase, thus inhibiting its action. Oxidation of hemoglobin to methemoglobin creates a large pool of CN^--binding protein which decreases $[CN^-]$ in mitochondria thus relieving the inhibition of cytochrome oxidase.

5. (1) — (a)
   (2) — (c)
   (3) — (b)

6. Nigericin exchanges K^+ ions for H^+ ions thereby discharging the proton concentration gradient across the inner mitochondrial membrane. Since this proton gradient is generated by electron transport and discharged by the ATP synthase-mediated formation of ATP, nigericin uncouples these two process. Valinomycin cannot do so since it does not transport protons. Its transport of K^+ will collapse the membrane potential but not change the proton concentration gradient thereby leaving ATP synthesis substantially undisturbed.

7. For transport of a proton from outside to inside,

$$\Delta G = RT\ln \frac{[H^+]_{in}}{[H^+]_{out}} + ZF\Delta\Psi = 2.3RT(pH_{out} - pH_{in}) + ZF\Delta\Psi$$

Since an ion is transported from the positive to the negative side of the membrane, $\Delta\Psi$ is negative.

$$\Delta G = 2.3 \times 8.314 \times 10^{-3} \times 298 \times (-1.4) + 1 \times 96.5 \times -0.06 = -7.97 - 5.79$$

$$\Delta G = -13.76 \text{ kJ} \bullet \text{mol}^{-1} \text{ H}^+ \text{ transported}$$

Since $\Delta G^{o'}$ for ATP synthesis is 30.5 kJ$\bullet$mol^{-1} and $30.5/13.76 = 2.22$, between 2 and 3 protons must be transported to provide the free energy required for the synthesis of 1 mole of ATP under standard biochemical conditions.

8. (a) The channel in F_o for translocating protons is blocked by F_1 so that protons may only be translocated when ATP is synthesized. Removal of F_1 unblocks this channel making submitochondrial particles permeable to protons.

(b) The presence of OSCP in the particles even after F_1 removal allows oligomycin to interact with F_o and block proton translocation.

9. Oligomycin inhibits F_oF_1-ATPase while CN^- inhibits cytochrome oxidase. Since electron

transport through cytochrome oxidase is coupled with ATP synthesis, both inhibitors inhibit the oxidative phosphorylation of pyruvate and succinate. Dinitrophenol uncouples oxidative phosphorylation so that substrate oxidation can occur in the absence of ATP synthesis. Oligomycin and CN^- inhibition may be distinguished by measuring O_2 uptake in the presence of dinitrophenol. Oligomycin does not inhibit O_2 uptake in the presence of dinitrophenol while CN^- does so.

10. The oxidation of glucose is an exothermic process, releasing the same amount of heat (enthalpy) independent of path. (Recall that enthalpy is the amount of heat released at constant pressure).

11. Atractyloside inhibits the ATP/ADP translocase, inhibiting mitochondrial oxidative phosphorylation by decreasing the amount of ADP available for phosphorylation.

12. Death is essentially an irreversible loss of order. On dying, cells loose their order on the molecular level by loosing their ion gradients, enzymatically digesting their macromolecular components, breaking down their membranes, *etc.* Thus, although cells and the organisms they comprise appear to change little on dying, the microscopic changes which occur are profound and cannot be reversed by simply "curing" the condition that caused death.

Chapter 21
OTHER PATHWAYS OF CARBOHYDRATE METABOLISM

1. The formation of G6P from glucose via the hexokinase reaction requires the expenditure of 1 ATP. The pentose phosphate pathway generates 2NADPH = 6ATP for each CO_2 released. The resulting Ru5P can be reconverted to G6P through the remaining reactions of the pentose phosphate pathway and gluconeogenesis (F6P $\rightarrow$ G6P and GAP $\rightarrow$ DHAP $\rightarrow$ FBP $\rightarrow$ F6P $\rightarrow$ G6P) without further input of energy. Thus, conversion of glucose to CO_2 via pentose phosphate + gluconeogenesis releases the energy of 6 x 6 – 1 = 35ATP.

   Glycolysis converts 1 glucose to 2pyruvate + 2ATP + 2NADH. The 2pyruvate are converted to 2acetyl-CoA + 2NADH by pyruvate dehydrogenase. The citric acid cycle converts 2acetyl-CoA to 6NADH + 2GTP + 2 $FADH_2$ + $4CO_2$. Hence, altogether this process generates 2ATP + 10NADH + 2 $FADH_2$ + 2GTP. Since each NADH = 3ATP, each $FADH_2$ = 2ATP, and each GTP = 1 ATP, this is equal to 2 + 3 x 10 + 2 x 2 + 2 = 38ATP.

   Consequently, oxidation via the pentose phosphate pathway is 35/38 x 100 = **92% as energetically efficient** as oxidation via glycolysis and the citric acid cycle.

2. Animals cannot carry out the net synthesis of glucose from acetyl-CoA (to which acetate is converted). ^{14}C-labeled acetyl-CoA nevertheless enters the citric acid cycle and thereby contributes ^{14}C to oxaloacetate (2 other C atoms are lost as CO_2 in the process so that there is no net synthesis of oxaloacetate). The labeled oxaloacetate may be converted to glucose through gluconeogenesis and subsequently taken up by muscle and converted to glycogen.

3. Inhibiting a particular step in the processing pathway prevents further processing and, therefore, the trimming precursors will accumulate so that they can be identified. By finding inhibitors for a series of sequential steps and identifying the resulting intermediates, the various steps in a pathway can be followed.

4. Through the intermediacy of the redox reactions the enzyme catalyzes,

$$NAD^+ + A_{red} \qquad NADH + A_{ox}$$

$$NADPH + A_{ox} \qquad NADP^+ + A_{red}$$

the NAD^+/NADH couple would be equilibrated with the $NADP^+$/NADPH couple (assuming the enzyme has sufficient catalytic activity to do so). This would then couple the pentose phosphate pathway to oxidative phosphorylation via the intermediacy of NADH. More importantly, there would be almost no NADPH available for biosynthesis since its reducing power would be transferred to NADH.

5. For

$$NADH + NADP^+ \qquad NAD^+ + NADPH$$

$$\Delta G = \Delta G^{o\prime} + RT \ln \frac{[NAD^+][NADPH]}{[NADH][NADP^+]}$$

But $\Delta G^{o\prime} = 0$.

Moreover, from Section 21-4, $[NAD^+]/[NADH] = 1000$ and $[NADP^+]/[NADPH] = 0.01$.

Thus,

$$\Delta G = RT \ln (1000/0.01) = 8.314 \text{ J} \bullet \text{K}^{-1} \text{ mol}^{-1} \times (273 + 37) \text{ K} \times \ln(10^5)$$

$$= 8.314 \times 310 \times 11.5 \text{ J} \bullet \text{mol}^{-1} = \textbf{29.6 kJ} \bullet \textbf{mol}^{-1}$$

6. Following the reactions of the pentose phosphate cycle (Figure 21-22):

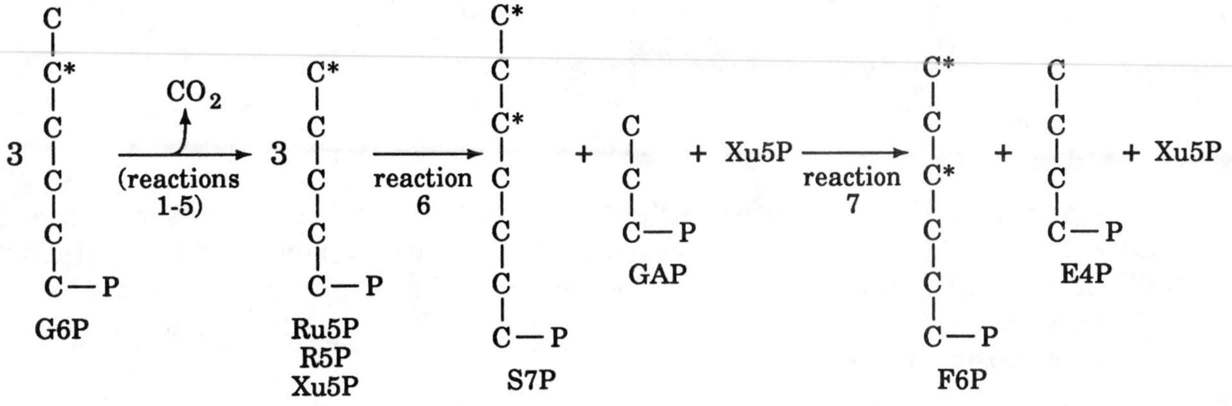

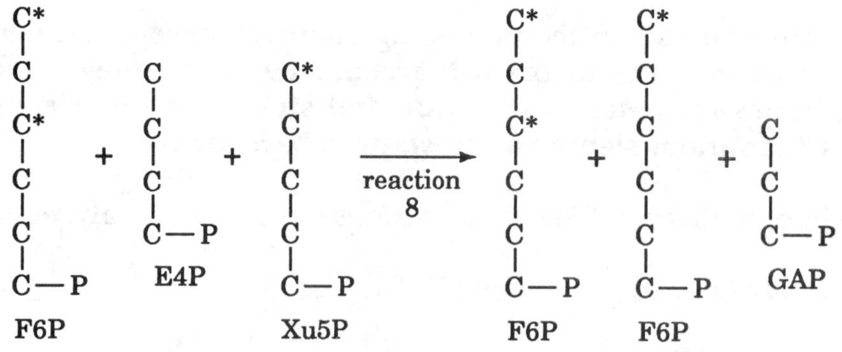

Thus, half the F6P product will be labeled at C(1) and C(3), the remainder of the F6P will be labeled at C(1), and none of the GAP will be labeled.

After gluconeogenesis, the 2GAP are converted to unlabeled G6P and each F6P is converted to a correspondingly labeled G6P. Then, after a passage through Reactions 1 to 3 of the pentose phosphate pathway, the two labeled C(1)'s of the G6P derived from F6P are lost as $^{14}CO_2$. This leaves a single Ru5P labeled in its C(2) position:

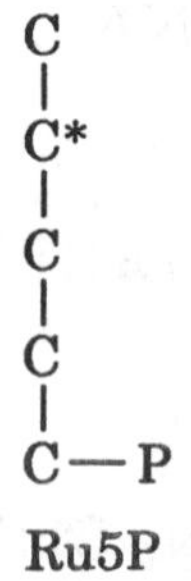

Ru5P

If this labeled Ru5P is converted to R5P (a 1/3 chance), reactions 6 and 7 will convert it to F6P labeled at C(3):

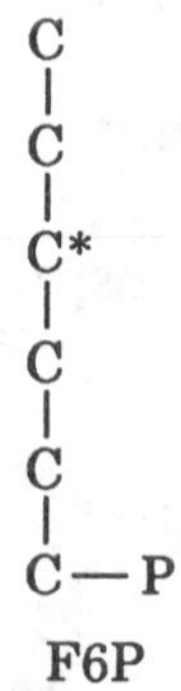

F6P

If the labeled Ru5P is converted to Xu5P (2/3 chance), half of it will react in reactions 6 and 7 to form C(2)-labeled FGP. The remainder of the C(2)-labeled Xu5P will react with E4P in reaction 8 to also form C(2)-labeled F6P. Thus, after 2 rounds of the pentose phosphate cycle, **2/3 of the original label will be in the CO_2, 1/9 at position C(3) of FGP, and 2/9 at position C(2) of F6P.**

7. In glycolysis, both C(1) and C(6) will be converted to CO_2 at the same rate. In the pentose phosphate pathway, C(1) is converted to CO_2 in one cycle of the pathway but all of the C(6) label is still associated with F6P and GAP. After a round of gluconeogenesis, 5/6 of this latter label (all of that in F6P and half of that in GAP) is associated with C(6) of G6P whereas 1/6 of the label (half of that in GAP) is at C(1). In a second round of the pentose phosphate pathway, the only labeled CO_2 to be generated will originate from this labeled C(1). Thus, in the pentose phosphate pathway, the C(6) position of glucose is converted to CO_2 at 1/6 the rate of the C(1) position (in a steady state process). Consequently, by comparing the rates of conversion of evolution of $^{14}CO_2$ from glucose labeled with ^{14}C at C(1) with that so-labeled at C(6), the relative rates of glycolysis *vs* that of the pentose phosphate pathway in a tissue may be determined.

8. It seems unlikely that any drug will be found to be absolutely safe, that is, have no atypical side effects. This is because some small fraction of the population will have a genetic predisposition to react atypically with almost any substance.

Chapter 22
PHOTOSYNTHESIS

1. Chlorophyll absorbs red and blue light and therefore transmits green light, the color we see.

2. (1)   P870 is analogous to P700 of PSI and P680 of PSII

   (2)   The BChl $a \rightarrow Q_{pool}$ chain on the reducing side of P870 resembles the Chl $a \rightarrow Q_{pool}$ chain on the reducing side of P680 of PSII. The analogous chain of PSI begins with A_0 and A_1, Chl a and phylloquinone molecules that respectively resemble the BChl a and menaquinone molecules of the bacterial system. However, the PSI chain then continues with ferredoxins as carriers, which is quite unlike the bacterial system.

   (3)   The cytochrome b-Fe–S electron transport chain in the bacterial system resembles, at least in function, the cytochrome b_6–f complex of PSII. PSI has no analogous system. However, PSI can return electrons to the cytochrome b_6–f complex in a cyclic electron transport process like that of the bacterial system.

3. Antimycin inhibits complex III ($CoQH_2$–cytochrome c reductase) in mitochondria (Section 20-2B). The analogous portion of the photosynthetic electron transport chain is the cytochrome b_6–f complex of PSII. This is the site of action of antimycin A in chloroplasts.

4. $E = hc/\lambda$

$$\lambda = 680 \text{ nm} = 680 \times 10^{-9} \text{ m}$$

$$E \text{ (per einstein)} = Nhc/\lambda$$

$$= 6.02 \times 10^{23} \text{ mol}^{-1} \times 6.626 \times 10^{-34} \text{ J} \bullet \text{sec} \times 2.998 \times 10^8 \text{ m} \bullet \text{sec}^{-1}/680 \times 10^{-9} \text{ m}$$

$$= 176 \text{ kJ} \bullet \text{mol}^{-1}$$

For every molecule of O_2 evolved by PSII, 4 electrons must be translocated and hence 4 photons absorbed. This results in the translocation of 12 protons and hence in the production of $12/3 = 4$ ATP.

PSI, in noncyclic electron transport, generates 2NADPH (each of which is equivalent to 3ATP) for every 4 electrons transported and thus 4 photons absorbed. In cyclic electron transport, which is independent of the action of PSII, each photon absorbed drives the translocation of 2 protons and thus the synthesis of 2/3 ATP.

In total, noncyclic transport produces the equivalent of $4 + 6 = 10$ ATP per 8 photons.

$$10 \text{ ATP} \times 59 \text{ kJ/mol ATP} = 590 \text{ kJ}$$

$$8 \text{ photons} \times 176 \text{ kJ/mol photons} = 1408 \text{ kJ}$$

$$\text{Efficiency} = 590 \text{ kJ} \times 100/1408 \text{ kJ} = \textbf{42\%}$$

For cyclic transport,

$$\text{Efficiency} = \frac{2/3 \times 59 \text{ kJ} \bullet \text{mol}^{-1} \times 100}{176 \text{ kJ} \bullet \text{mol}^{-1}} = \textbf{22\%}$$

500 nm light is $680/500 = 1.36$ times more energetic per photon than 680 nm light. Thus, for noncyclic transport, the efficiency with 500 nm light is $42\%/1.36 = \textbf{31\%}$ and with cyclic transport it is $22\%/1.36 = \textbf{16\%}$.

5. $$\text{ADP} + P_i \rightleftharpoons \text{ATP} + H_2O$$

$$\Delta G = \Delta G^{\circ\prime} + RT\ln\frac{[\text{ATP}]}{[\text{ADP}][P_i]}$$

$$\frac{[\text{ATP}]}{[\text{ADP}][P_i]} = 10^3 \text{ at } 25°C.$$

Table 15-3 indicates the $\Delta G^{\circ\prime}$ for this process is $+30.5 \text{ kJ} \bullet \text{mol}^{-1}$. Consequently,

$$\Delta G = 30.5 \times 10^3 \text{ J} \bullet \text{mol}^{-1} + 8.314 \text{ J} \bullet \text{K}^{-1} \text{ mol}^{-1} \times (273 + 25) \text{ K} \ln 10^3$$

$$= 30{,}500 + 17{,}100 = 47{,}600 \; J{\bullet}mol^{-1}$$

A proton concentration gradient with this free energy is needed to generate ATP.

$$3H^+(in) \rightarrow 3H^+(out)$$

where (*out*) indicates the stroma and (*in*) indicates the thylakoid lumen.

$$\Delta G = \Delta G^{\circ\prime} + RT \ln \frac{[H^+(out)]^3}{[H^+(in)]^3} = -47{,}600 \; J{\bullet}mol^{-1}$$

But $\Delta G^{\circ\prime} = 0$ since the same components are reactants and products (the membrane potential in chloroplasts is nearly zero and hence need not be taken into account).

$$\frac{[H^+(out)]}{[H^+(in)]} = e^{-47{,}600 \; J{\bullet}mol^{-1} \; /[3 \times 8.314 J{\bullet}K^{-1} \; mol^{-1} \times (273 + 25) \; K]} = 0.00165$$

$$\Delta pH = -\log [H^+(out)] - (-\log [H^+(in)]) = -\log\left(\frac{[H^+(out)]}{[H^+(in)]}\right) = -\log 0.00165$$

$\Delta pH = \textbf{2.78}$, that is, the stroma must be 2.78 pH units higher than the thylakoid lumen to generate ATP under the conditions given.

6. RuBP after one round of labeling has the average pattern (see Figure 22-23):

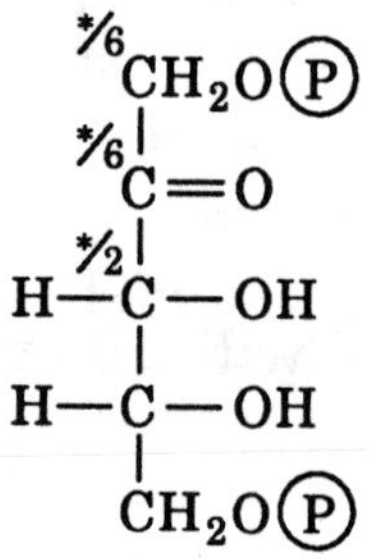

$$\begin{array}{l} {}^{*/6}CH_2O\,\text{\textcircled{P}} \\ {}^{*/6}|\\ C{=}O \\ {}^{*/2}| \\ H{-}C{-}OH \\ | \\ H{-}C{-}OH \\ | \\ CH_2O\,\text{\textcircled{P}} \end{array}$$

where */6, for example, indicates the atom to be $1/6$ as labeled as the original $^{14}CO_2 = $ $^*CO_2$.

The second round of reaction with $^*CO_2$ yields:

$$^{*}\!/_6\;CH_2O\,\textcircled{P}$$
$$H-\!\!\!\overset{*/_6}{C}-OH$$
$$^{*}CO_2^-$$
$$+$$
$$^{*}\!/_2\;CO_2^-$$
$$H-C-OH$$
$$CH_2O\,\textcircled{P}$$

3PG

$\left.\begin{array}{c}\\\\\\\\\\\\\\\end{array}\right\}$ average $=$

$$^{3}\!/_4^{*}\;CO_2^-$$
$$H-\!\!\!\overset{*/_{12}}{C}-OH$$
$$^{*}\!/_{12}\;CH_2O\,\textcircled{P}$$

3PG $\longrightarrow$

$$^{3}\!/_4^{*}\;CHO$$
$$H-\!\!\!\overset{*/_{12}}{C}-OH$$
$$^{*}\!/_{12}\;CH_2O\,\textcircled{P}$$

GAP

$$^{*}\!/_{12}\;CH_2O\,\textcircled{P}$$
$$^{*}\!/_{12}\;C=O$$
$$^{3}\!/_4^{*}\;CH_2OH$$

DHAP

$$^{*}\!/_{12}\;CH_2OH$$
$$^{*}\!/_{12}\;C=O$$
$$HO-\!\!\!\overset{3/_4^{*}}{C}-H$$
$$H-\!\!\!\overset{3/_4^{*}}{C}-OH$$
$$H-\!\!\!\overset{*/_{12}}{C}-OH$$
$$^{*}\!/_{12}\;CH_2O\,\textcircled{P}$$

F6P

GAP $\longleftarrow$ ... $\xleftarrow{\;GAP\;}$

$$^{*}\!/_{12}\;CH_2OH$$
$$^{*}\!/_{12}\;C=O$$
$$HO-\!\!\!\overset{3/_4^{*}}{C}-H$$
$$H-\!\!\!\overset{*/_{12}}{C}-OH$$
$$^{*}\!/_{12}\;CH_2O\,\textcircled{P}$$

Xu5P

$+$

$$^{3}\!/_4^{*}\;CHO$$
$$H-\!\!\!\overset{3/_4^{*}}{C}-OH$$
$$H-\!\!\!\overset{*/_{12}}{C}-OH$$
$$^{*}\!/_{12}\;CH_2O\,\textcircled{P}$$

E4P

$\xrightarrow{\;DHAP\;}$

$$^{*}\!/_{12}\;CH_2OH$$
$$^{*}\!/_{12}\;C=O$$
$$HO-\!\!\!\overset{3/_4^{*}}{C}-H$$
$$H-\!\!\!\overset{3/_4^{*}}{C}-OH$$
$$H-\!\!\!\overset{3/_4^{*}}{C}-OH$$
$$H-\!\!\!\overset{*/_{12}}{C}-OH$$
$$^{*}\!/_{12}\;CH_2O\,\textcircled{P}$$

S7P

$\xrightarrow{\;GAP\;}$

$$^{3}\!/_4^{*}\;CHO$$
$$H-\!\!\!\overset{3/_4^{*}}{C}-OH$$
$$H-\!\!\!\overset{3/_4^{*}}{C}-OH$$
$$H-\!\!\!\overset{*/_{12}}{C}-OH$$
$$^{*}\!/_{12}\;CH_2O\,\textcircled{P}$$

R5P

$+$

$$^{*}\!/_{12}\;CH_2OH$$
$$^{*}\!/_{12}\;C=O$$
$$HO-\!\!\!\overset{3/_4^{*}}{C}-H$$
$$H-\!\!\!\overset{*/_{12}}{C}-OH$$
$$^{*}\!/_{12}\;CH_2O\,\textcircled{P}$$

Xu5P

so that the average RuBP labeling after the second round is:

$$^{11}\!/_{36}^{*}\;CH_2OH$$
$$^{11}\!/_{36}^{*}\;C=O$$
$$H-\!\!\!\overset{3/_4^{*}}{C}-OH$$
$$H-\!\!\!\overset{*/_{12}}{C}-OH$$
$$^{*}\!/_{12}\;CH_2O\,\textcircled{P}$$

7. Shortly after the light is turned off, the ATP and NADPH concentrations drop as these substances are used up in the Calvin cycle (Figure 22-23) without replacement by the light reactions. **3PG builds up** because it cannot pass through the phosphoglycerate kinase reaction in the absence of ATP. However, **Ru5P drops** because it is consumed by the ribulose bisphosphate carboxylase reaction (which requires neither ATP nor NADPH) but its replenishment is blocked by the lack of ATP in the phosphoribulokinase reaction.

8. The Calvin cycle (Figure 22-23) requires the input of 9ATP + 6NADPH to synthesize 1 GAP. This can be converted to G1P without further energy input. Conversion of G1P to starch requires the energy of 2ATP for a total energy input of 2(9ATP + 6NADPH) + 2 ATP = 20ATP + 12NADPH = 56ATP.

   The starch → G1P reaction requires no energy input. The breakdown of G1P to 2pyruvate yields 3ATP + 2NADH (the glycolysis of G1P nets 3ATP because the hexokinase reaction has been bypassed) while conversion of 2pyruvate to 2acetyl-CoA yields 2NADH. Conversion of 2acetyl-CoA to CO_2 + H_2O via the citric acid cycle (Section 19-1A) yields 6NADH + 2$FADH_2$ + 2GTP (= 2ATP). Thus, the oxidation of starch to CO_2 + H_2O yields 10NADH + 2$FADH_2$ + 5ATP per glucose unit. Since the oxidative phosphorylation of NADH and $FADH_2$ respectively yield 3 and 2ATP, the oxidation of starch yields 3 x 10 + 2 x 2 + 5 = 39ATP per glucose unit. The energetic efficiency of the system therefore is:

$$\text{Efficiency} = 39/56 \times 100 = \textbf{70\%}$$

9. The partial pressure of CO_2 at which a C_4 plant can photosynthesize is much lower than that of a C_3 plant. Moreover, at very low partial pressures of CO_2, the C_3 plant reverses the effects of photosynthesis by photorespiration. In the sealed box, the C_4 plant maintains the CO_2 partial pressure so low that the C_3 plant wastes away through photorespiration. In effect, the C_4 plant devours the C_3 plant.

10. The plants in question store CO_2 by CAM. During the course of the night, CO_2 reacts with PEP to form malic acid. By morning, so much of this acid has been generated that the plant's leaves have a sour taste. During daylight hours, however, the malate is converted to pyruvate + CO_2 and the CO_2 is fixed by photosynthesis. The leaves therefore become less acid and hence tasteless until late in the day when they may have consumed all their malate so as to become slightly basic, that is, bitter.

Chapter 23
LIPID METABOLISM

1. Phospholipase A_2 catalyzes the hydrolysis of the fatty acid residue at C(2) of a phospholipid to yield the corresponding lysophospholipids. These latter compounds are powerful detergents and therefore disrupt biological membranes, apparently to a greater extent than the direct action of the phospholipase A_2 itself.

2. Jamaican vomiting sickness results from the MCPA-CoA-inhibition of acyl-CoA dehydrogenase thereby preventing fatty acid breakdown. The body is therefore forced to depend only on its limited carbohydrate reserves for energy metabolism. This consists mainly of glycogen in the liver and muscle which is depleted in a matter of hours by normal metabolic energy requirements.

3. Tripalmitoyl glycerol ($C_{51}H_{98}O_6$, 806 D) is hydrolyzed by "hormone-sensitive" triacylglycerol lipase to three molecules of palmitic acid and one of glycerol. Complete oxidation of each palmitic acid yields a net of 129ATP (Section 23-3C). The glycerol is oxidized to dihydroxyacetone phosphate (DHAP) with the expenditure of 1 ATP and a yield of 1 NADH (= 3ATP) for a net gain of 2ATP. The DHAP, upon glycolysis, yields pyruvate together with 2ATP + 1 NADH (= 3ATP) for a net of 5ATP. The pyruvate is metabolized to CO_2 and H_2O via pyruvate dehydrogenase (Section 19-2A) and the citric acid cycle (Section 19-1) yielding 1 GTP (= 1 ATP) + 4NADH (= 12ATP) + 1 $FADH_2$ (= 2ATP) for a total of 15ATP. Thus, the amount of ATP generated by the complete oxidation of tripalmitoyl glycerol is 3 x 129 + 2 + 5 + 15 = 409ATP. This yields 409 moles ATP/806 g tripalmitoyl glycerol = **0.51 mol ATP/g tripalmitoyl glycerol.**

Glycogen [$(C_6H_{10}O_5)_n$; 162 D/glucose unit] is broken down by glycogenolysis to 90% glucose-1-phosphate (G1P) and 10% glucose (Section 17-1). G1P is isomerized to G6P. The complete oxidation of glucose via glycolysis, the citric acid cycle, and oxidative phosphorylation yields 38ATP (Section 20-4C) whereas that of G6P yields 39ATP. Hence, oxidation of glycogen yields $(0.9 \times 39 + 0.1 \times 38)$ mol of ATP/(162 g glycogen + 2 × 162 g H_2O) = **0.08 mol ATP/g hydrated glycogen**. The oxidation of tripalmitoyl glycerol therefore yields 0.51/0.08 = **6.3 times more ATP than an equal weight of hydrated glycogen**.

4. (a) The palmitic acid forms palmitoyl-CoA in a reaction catalyzed by acyl-CoA synthase and then is broken down via β oxidation to yield acetyl-CoA. If the carboxyl C of palmitic acid is ^{14}C-labeled, the thioester C of the product acetyl-CoA will likewise be ^{14}C-labeled. If palmitic acid is synthesized from this labeled acetyl-CoA, then all of the palmitic acid's odd numbered C atoms will be labeled:

$$CH_3 \!-\!(^{14}CH_2 \!-\! CH_2)_{\overline{7}} \!-\! {}^{14}COOH$$

(b) The labeled acetyl-CoA can pass through the citric acid cycle yielding labeled oxaloacetate (Figure 19-1) which, via gluconeogenesis, yields labeled glucose-6-phosphate (Figure 21-7). This, in turn, is converted to labeled glycogen via reactions described in Section 17-2 (notably the glycogen synthase reaction). However, there is no animal pathway that mediates the net synthesis of oxaloacetate from acetyl-CoA. Rather, the citric acid cycle oxidizes an acetyl-CoA equivalent while incorporating another acetyl-CoA equivalent into oxaloacetate. Nevertheless, acetyl-CoA from fatty acid breakdown can "exchange" via gluconeogenesis and glycogen synthesis with acetyl-CoA from glycogen as generated by glycogenolysis and glycolysis yielding labeled glycogen without net glycogen synthesis.

5. (a) Stearic acid (18:0; the saturated analog of α-linolenic acid) β oxidation would yield 9acetyl-CoA (= 9GTP + 27NADH + 9FADH$_2$) + 8FADH$_2$ + 8NADH – 2ATP. Since 1 GTP = 1 ATP, 1 FADH$_2$ = 2ATP and 1 NADH = 3ATP, this is equivalent to 146ATP. However, a double bond at an odd-numbered C (of which α-linolenic has two) is converted, via the enoyl-CoA isomerase reaction, to a cis-β-double bond. Hence, the acyl-CoA dehydrogenase reaction, which would otherwise form this bond as part of β oxidation, is unnecessary; that is, one less FADH$_2$ (= 2ATP) will be generated per odd numbered double bond. A double bond at an even numbered C (α-linolenic acid has one) must be reduced by NADPH (= 3ATP). α-Linolenic acid oxidation therefore yields 146 – 2 × 2 – 3 = **139ATP**.

(b) Margaric acid oxidation yields propionyl-CoA + 7 acetyl CoA (= 7GTP + 21NADH + 7FADH$_2$) + 6FADH$_2$ + 6NADH – 2ATP = 1 propionyl-CoA + 112ATP. Propionyl-CoA is converted to succinyl-CoA with the expenditure of 1 ATP (Figure 23-14). Succinyl-CoA is converted to oxaloacetate with the generation of 1 GTP + 1 FADH$_2$ + 1 NADH = 6ATP in the citric acid cycle. The oxaloacetate is converted to pyruvate with no net ATP expenditure by the sequential actions of PEP carboxykinase and pyruvate kinase (Figure 21-7). The pyruvate is oxidized via the citric acid cycle (Figure 19-1) yielding 1 GTP + 1

FADH$_2$ + 4NADH = 15ATP. Margaric acid therefore yields 112 − 1 + 6 + 15 = **132ATP.**

α-Linolenic acid yields 139/18 = 7.72 ATP/C whereas margaric acid yields 132/17 = 7.76 ATP/C. Hence, **margaric acid yields slightly more ATP/C than does α-linolenic acid.**

6. The C—Co bond of dioldehydrase's coenzyme B$_{12}$ apparently undergoes homolytic fission to yield a 5′-adenosyl free radical (Figure 23-18, Step 1). This reaction abstracts one of 1,2-propandiol's two C(1) H atoms and, following rearrangement of this substrate, returns an H atom to C(2) of the product, propionaldehyde (analogous to Reactions 2-4 of Figure 23-18). Upon abstraction of a ^{3}H atom by the 5′-adenosyl free radical, the resulting 5′-methyl group freely rotates about its C(4′)—C(5′) bond thereby randomizing the label through all of its 3 hydrogen positions. One of the these H atoms is returned to the substrate leaving the other two C(5′) positions labeled.

   If the enzyme is supplied with [5′-^{3}H]-deoxyadenosylcobalamin and unlabeled 1,2-propandiol, 1/3 of the label will be transferred to the product in each reaction cycle yielding [2-^{3}H]-propionaldehyde. The C(5′) group of the product 5′-adenosylcobalamin would be uniformly ^{3}H-labeled but at 2/3 of the level it had before the reaction cycle.

7. Breaking down palmitic acid to acetyl-CoA requires one cycle of the acyl-CoA synthetase reaction to yield palmitoyl-CoA (uses 1 ATP) and 7 cycles of β oxidation to yield 8acetyl-CoA [yielding 1 FADH$_2$ (= 2ATP) and 1 NADH (= 3ATP) per cycle]. The process therefore produces 7 x (2 + 3) −1 = 34ATP.

   Synthesis of palmitic acid from acetyl-CoA requires seven cycles of the fatty acid synthase reaction which consumes 2NAPH per cycle = 7 x 6 = 42ATP. In addition, the formation of the 7malonyl-CoA via the acetyl-CoA carboxylase reaction consumes 7ATP for a total of 42 + 7 = 49ATP.

   Thus, the energetic price of breaking down and resynthesizing palmitic acid is 49 − 34 = **15ATP.**

8. The formation of HMG-CoA from acetyl-CoA (Figure 23-20) requires no ATP input (the reaction is driven by thioester hydrolysis). Its transformation to isopentenyl pyrophosphate requires 2NADH (= 6ATP) + 3ATP = 9ATP (Figures 23-37). Six isopentenyl pyrophosphates are needed to synthesize cholesterol so that this process requires 6 x 9 = 54ATP. The two prenyl transferase reactions require no ATP input but the squalene synthase reaction utilizes 1 NADPH = 3ATP (Figure 23-40). The squalene epoxidase reaction also requires 1 NADPH = 3ATP (Figure 23-44). The squalene oxidocyclase reaction requires no ATP input (Figure 23-45) but the conversion of its product, lanosterol, to cholesterol requires a net of 13NAD(P)H = 39ATP (Figure 23-46). Hence, a total of 54 + 3 + 3 + 39 = **99 ATP equivalents** are required to synthesize cholesterol from acetyl-CoA.

9. (a) HMG-CoA's C(5) carbonyl group is lost as CO$_2$ in the pyrophosphomevalonate decarboxylase reaction forming isopentenyl pyrophosphate. Hence, **the cholesterol**

made from [5-^{14}C]HMG-CoA would be unlabeled.

(b) [1-^{14}C]HMG-CoA yields [1-^{14}C]isopentenyl pyrophosphate and [1-^{14}C]dimethylallyl pyrophosphate. Thus, through the prenyl transferase and squalene synthase reactions (Figure 23-40), the squalene synthesized would have the labeling pattern:

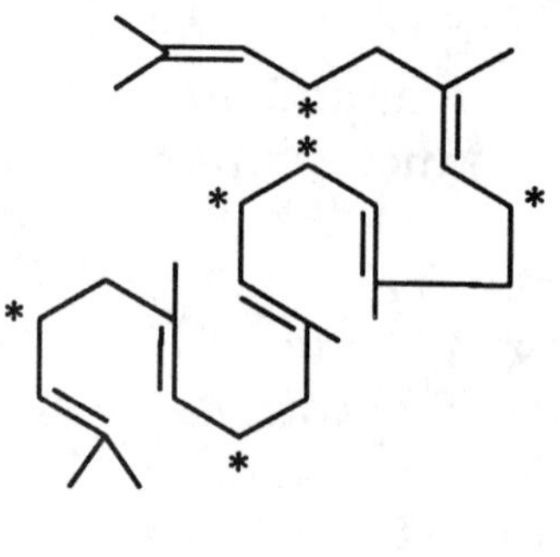

The cyclization reaction to form lanosterol gives the analogous labeling pattern (Figure 23-45) because none of the migrating methyl groups are labeled. The subsequent processing to yield cholesterol (Figure 23-46) alters the carbon skeleton only by removing three of the unlabeled methyl groups yielding cholesterol labeled as indicated:

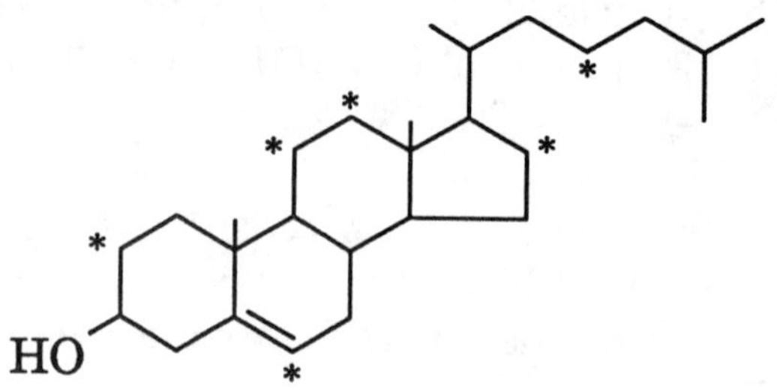

that is, at C(2), C(6), C(11), C(12), C(16) and C(23).

10. The patient is suffering from a deficiency of lipoprotein lipase (an inherited disease) which normally functions to hydrolyze the triacylglycerols of chylomicrons and VLDL. The symptoms of the disease may be minimized by maintaining a low fat diet since chylomicrons are intestinally packaged dietary lipids.

11. Linoleic acid is an eiscosanoid precursor that animals cannot synthesize. It must therefore be supplied in the diet in order to maintain intercellular communications via prostaglandins, thromboxanes, and leukotrienes and to provide an essential component of the skin's water barrier. Animal cells in culture do not need to maintain intercellular communications nor do they require a water barrier. Consequently, they do not require linoleic acid.

Chapter 24
AMINO ACID METABOLISM

1. The urea cycle functions to transform excess nitrogen to an excretable form, urea. This excess nitrogen arises through the breakdown of proteins. If there is a deficiency in a urea cycle enzyme, the preceeding urea cycle intermediates may build up to a toxic level. Hence, a low protein diet, which minimizes excess nitrogen, reduces the concentrations of these toxic substances.

2. The nitrogen in a high protein diet must be disposed of as urea. Since urea is excreted in a relatively dilute aqueous solution (urine), a high protein diet must be accompanied by sufficient water intake to form the required amount of urine.

3. 40 g of urea contains (2×14 g N•mol^{-1}/60 g urea•mol^{-1}) $\times$ 40 g urea = 18.7 g N, which corresponds to 18.7 g/0.16 = 116.7 g protein. This amount of protein provides 116.7 g $\times$ 18 kJ/g = 2100 kJ of energy or (2100/10,000) $\times$ 100 = **21% of the student's daily energy requirement.**

4. In the absence of carbohydrate intake, all citric acid cycle intermediates must be derived from the glucogenic amino acids obtained in the diet or from protein degradation. The bad breath, which is due to the exhalation of acetone, a ketone body, indicates that there are, nevertheless, insufficient citric acid cycle intermediates to metabolize all the acetyl-CoA produced by lipid and ketogenic amino acid degradation (a condition known as ketosis; Section 25-3B). The loss of ketone bodies via acetone exhalation makes lipid and amino acid metabolism less efficient than it would be with adequate supplies of carbohydrates. However, the intake of large quantities of fat and amino acids would

inhibit the consumption of body fat, even in the absence of dietary carbohydrate. Weight loss is therefore unlikely unless fat and protein intake are also curtailed.

5. Aspartame is digested to yield phenylalanine which, in phenylketonurics, is degraded to the toxic (at least in children) phenylpyruvate as a consequence of a defective enzyme in their phenylalanine degradation pathway.

6. According to the pathway diagrammed in Figure 24-28, PBG, labelled as indicated in Fig. 24-27, yields uroporphyrinogen III labeled as follows:

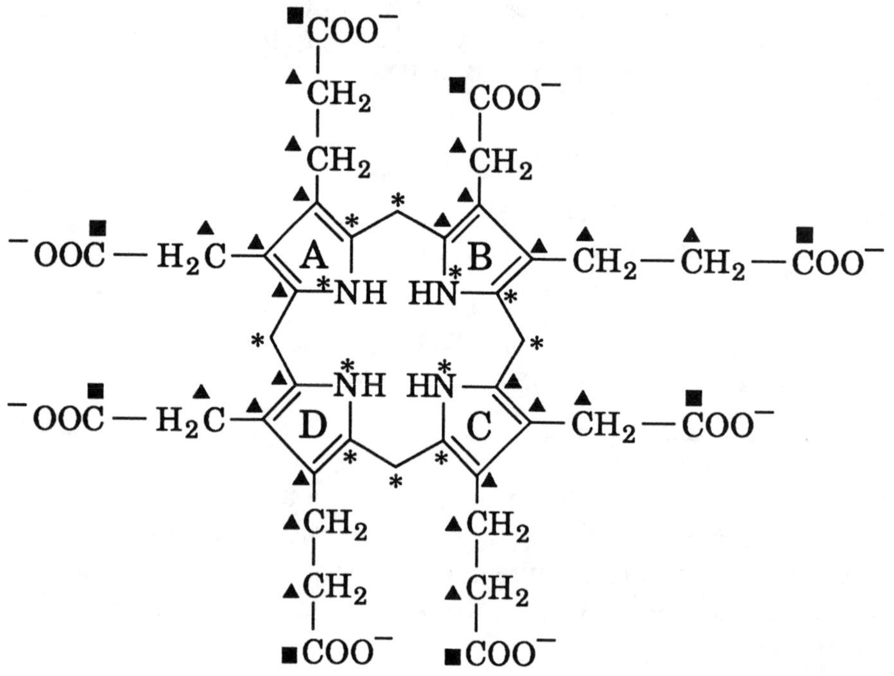

Decarboxylation of the acetate residues to methyl groups and oxidative decarboxylation of the A and B ring propyl groups to yield vinyl groups then yields the labeling pattern indicated in Figure 24-24.

7. Many substances are detoxified in the liver through the action of cytochrome P_{450}. The ingestion of drugs and chemicals therefore stimulates heme biosynthesis so as to form the required amounts of cytochrome P_{450}. This process yields porphyrins in the amounts that victims of acute intermitant porphyria cannot degrade thereby conferring the symptoms of this disease.

8. The pigment coloring skin and hair is melanin which is synthesized from tyrosine. When tyrosine is in short supply, as when dietary protein is not available, melanin cannot be synthesized in normal amounts so that the skin and hair become depigmented.

9. In the absence of uridylyl-removing enzyme activity, adenylyltransferase•P_{II} will be fully uridylylated since no mechanism exists for removing its uridylyl groups once they

are attached. Uridylylated adenylyltransferase•P_{II} adenylylates glutamine synthetase so as to put it in its more active form. Hence, an *E. coli* with defective uridylyl-removing enzyme will have a hyperactive glutamine synthetase and thus a higher than normal glutamine concentration. Reactions requiring glutamine will therefore be accelerated, thus depleting glutamate and the citric acid cycle intermediate α-ketoglutarate. Consequently, biosynthetic reactions requiring transamination, as well as energy metabolism, are suppressed.

10. *S*-Adenosylmethionine, an intermediate in methionine breakdown, is an important biological methylating agent. If methionine breakdown occurred via the reverse of its synthesis, this essential substance would not be synthesized.

11. Under anaerobic conditions, muscle generates its required ATP through glycolysis. This process generates pyruvate and NADH which can further react to form lactate and regenerate NAD^+ for further glycolysis. If the pyruvate is transaminated to alanine which is exported, then some other mode of NAD^+ regeneration must occur if the cell is to continue metabolizing. Numerous amino acid degradation reactions generate NH_3 which, in turn, reacts with α-ketoglutarate to form glutamate. In the process, NADH is oxidized to NAD^+. Thus, the glutamate dehydrogenase reaction generates the NAD^+ necessary to form pyruvate via glycolysis and also generates glutamate to be used in the transamination reaction with pyruvate to yield alanine and α-ketoglutarate.

12. The presence of both photosystem I and photosystem II in cyanobacteria evolves O_2 (Section 22-2C) which would poison nitrogenase. PSI, however, does not evolve O_2 and can operate in a cyclic manner so as to produce ATP, which is required in large amounts to power nitrogen fixation.

Chapter 25
ENERGY METABOLISM: INTEGRATION AND ORGAN SPECIALIZATION

1. The liver is the only organ capable of urea biosynthesis so that, upon liver failure, the ammonia level in the blood rises leading to ammonia toxicity. Severe hypoglycemia also ensues because the liver normally functions to maintain blood glucose levels between meals via gluconeogenesis. Many toxic substances will accumulate in the blood because the liver is unable to detoxify them.

2. Since athletes can sustain greater exertion before experiencing muscle fatigue than normal individuals, and since muscle fatigue results from a metabolically induced pH drop at the muscles, it is hypothesized that athletes' muscles are more heavily buffered than those of normal individuals.

3. During starvation, energy metabolism is largely shifted to the consumption of fats rather than amino acids derived from proteins. The excess ammonia resulting from amino acid metabolism is disposed of via the synthesis and secretion of urea. Hence, if only small amounts of amino acids are metabolized, little ammonia will be generated for urea synthesis.

4. Uncontrolled diabetics produce large amounts of ketone bodies, such as acetoacetate, that circulate in the blood. This compound easily decarboxylates to yield acetone.

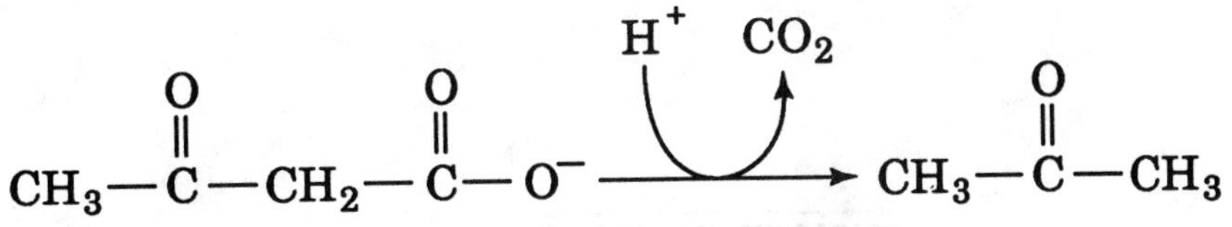

$$CH_3 - \overset{\displaystyle O}{\overset{\|}{C}} - CH_2 - \overset{\displaystyle O}{\overset{\|}{C}} - O^- \xrightarrow{\quad H^+ \quad CO_2 \quad} CH_3 - \overset{\displaystyle O}{\overset{\|}{C}} - CH_3$$

This volatile compound is, in part, released in the breath.

Chapter 26
NUCLEOTIDE METABOLISM

1. The reactions in which glutamine participates are Reactions 2 and 5 of the IMP pathway (Figure 26-3; amidophosphoribosyl transferase and FGAM synthetase), the GMP synthetase reaction of GMP synthesis (Figure 26-5), and Reaction 1 of pyrimidine biosynthesis (Figure 26-8; carbamoyl phosphate synthetase II). The intermediates that accumulate are therefore PRPP, FGAR and XMP (the reactants of the carbamoyl phosphate synthetase reaction do not accumulate).

2. In analogy with the mechanism in Figure 26-4,

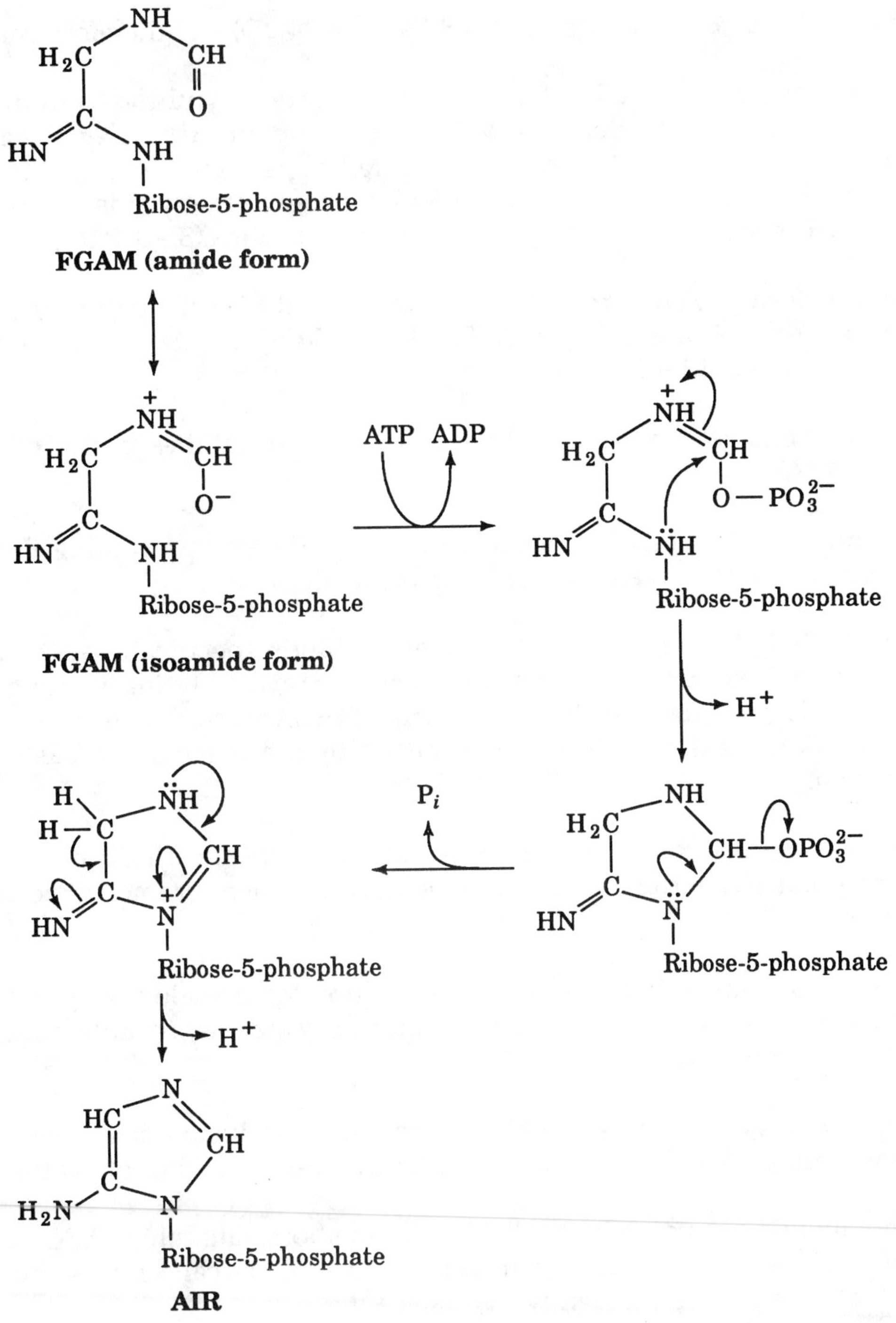

FGAM (amide form)

FGAM (isoamide form)

Ribose-5-phosphate

AIR

3. The synthesis of the hypoxanthine residue of IMP directly consumes 6ATP [the PP_i released in Reaction 2 of the pathway (Fig. 26-3) is energetically equivalent to an ATP]. In addition, we must recycle the other products of the pathway:

(1)   2Glutamate → 2glutamine requires 2ATP (Section 24-5A).

(2)   Fumarate → asparate yields 1 NADH (Section 24-2D) and therefore generates 3ATP.

(3)   The glycine taken up in Reaction 3 of the pathway is synthesized from serine (serine + NADH → 2glycine; Section 24-3B). The serine, in turn, is synthesized from the glycolytic intermediate 3-phosphoglycerate in a reaction yielding 1 NADH (Section 24-5A). The complete metabolism of 3-phosphoglycerate yields 16ATP (Chapters 16 and 19). The synthesis of one glycine therefore requires $(3 - 3 + 16)/2 = 8$ATP.

(4)   The $2N^{10}$-formyl-THF taken up in Reactions 4 and 10 of the pathway can be recycled at the expense of one serine (13ATP) while yielding one NADH (Sections 24-3B and 4D) for a total of $13 - 3 = 10$ATP.

Thus, the total energetic price of synthesizing the hypoxanthine residue of IMP is $6 + 2 - 3 + 8 + 10 = \textbf{23ATP}$.

4. Deoxyadenosine inhibits ribonucleotide reductase thereby preventing the synthesis of the deoxynucleotides cells require for DNA biosynthesis.

5. (a)   Tosyl-L-phenylalanine chloromethylketone binds specifically at the active site of chymotrypsin where it alkylates His 57, thereby irreversibly inactivating the enzyme. This reaction is not catalyzed by the enzyme, however, so that tosyl-L-phenylalanine chloromethylketone is an affinity label rather than a mechanism-based inhibitor of chymotrypsin.

(b)   Trimethoprim binds only to bacterial dihydrofolate reductase. It does not permanently inactivate the enzyme. Trimethoprim is therefore not a mechanism-based inhibitor.

(c)   The δ-lactone analog of $(NAG)_4$ is a transition state analog of lysozyme's natural substrate. Although it binds quite specifically to lysozyme, it undergoes no reaction. Hence, it is not a mechanism-based inhibitor.

(d)   Allopurinol is oxidized by xanthine oxidase to a product which irreversibly binds to the enzyme. Allopurinol is therefore a mechanism-based inhibitor of xanthine oxidase.

6. Chemotherapeutic agents such as FdUMP and methotrexate inhibit DNA synthesis and hence cell proliferation. Rapidly proliferating cells, such as cancer cells and those of hair follicles, therefore die. Consequently, hair falls out.

7. Cells defective in thymidylate synthase survive in the medium because it contains the thymidine they are unable to make. Normal cells, however, continue to make their own thymidine and thus convert their limited supply of folate cofactors to dihydrofolate. The methotrexate inhibits dihydrofolate reductase thereby preventing the regeneration of the tetrahydrofolate required in other reactions of nucleotide biosynthesis as well as several

reactions of amino acid biosynthesis. The cells then die from lack of these essential substances.

8. FdUMP inhibits thymidylate synthase thereby blocking TTP synthesis. Methotrexate inhibits dihydrofolate reductase so as to prevent regeneration of dihydrofolate to tetrahydrofolate, which is needed in other reactions. When these substances are taken together, the FdUMP slows the conversion of tetrahydrofolate to dihydrofolate and hence to some extent counteracts the effect of methotrexate. Similarly, since FdUMP can only inactivate thymidylate synthase in the presence of N^5,N^{10}-methylenetetrahydrofolate, the presence of methotrexate, which slows the regeneration of this cofactor, thereby slows the inactivation of thymidylate synthase.

9. Meat contains relatively large quantities of nucleic acids which are broken down to yield, among other products, uric acid. This additional uric acid may increase the uric acid concentration beyond the solubility limit in some otherwise gout-free individuals, thereby precipitating a gout attack.

10. Glycine is incorporated in the purine ring during its *de novo* synthesis. If the gout is due to overproduction of purines, the excreted uric acid should contain a high proportion of ^{15}N. If the gout is due to impaired excretion of uric acid, much of the excess uric acid will arise from the degradation of ingested purines and hence be ^{15}N-free. The fraction of excreted uric acid containing ^{15}N will therefore be relatively low.

11. 6-Mercaptopurine, being a hypoxanthine analog, is oxidized to 2-oxo-6-mercaptopurine by xanthine oxidase. This reacion largely inactivates this chemotherapeutic agent. The presence of allopurinol inactivates xanthine oxidase thereby increasing the effective concentration of 6-mercaptopurine.

Chapter 27

DNA: THE VEHICLE
OF INHERITANCE

1. The testcross for peas involves a cross between the F_1 heterozygote Yy and the parent that is homozygous for the recessive trait yy.

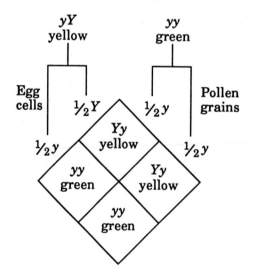

Half the progeny of this testcross will be green (yy) and half will be yellow (Yy). For snapdragons, in which r and R are codominant, this would be, using the white parent in the testcross:

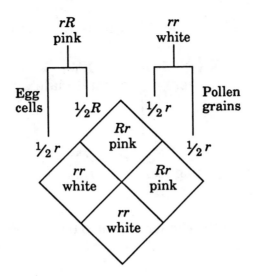

which yields progeny half of which are pink (*Rr*) and half of which are white (*rr*).

2. **The father of child 1 is male 1** because any child of male 2, being N ($L^N L^N$ genotype), would have to be either MN or N in phenotype, whereas male 1, being MN ($L^M L^N$ genotype) could have a type M child with a type M mother.

   The father of child 2 could be either male 1 or male 2 according to the data given since either male could have children of the blood type given with the mother.

   The father of child 3 is male 2 because male 2, having type AB ($I^A I^B$ genotype), could have a type AB child with the type B mother ($I^B I^B$ or $I^B i$ genotype) whereas male 1, who is also type B, could only have a type B or possibly type O child with the mother.

3. The color blind allele is sex-linked; that is, it is carried on the X chromosome. Thus, the children would have the genotypes:

$$XX \quad \times \quad X^{CB} Y \quad (X^{CB} \text{ contains the color blind allele})$$
mother \qquad father

$$\downarrow$$

$$XX^{CB} \quad + \quad XX^{CB} \quad + \quad XY \quad + \quad XY$$
daughter (carrier) \quad daughter (carrier) \quad son \quad son

None of the sons but all of the daughters will receive the color blind allele. The daughters with the color blind allele will be phenotypically normal because the color blind trait is recessive. However, they are "carriers" of the allele.

For the second generation, only the offspring of the carrier need be considered:

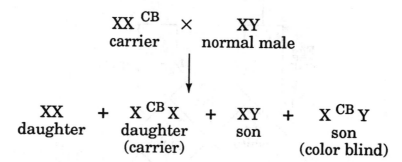

XX^{CB} × XY
carrier   normal male

XX + X^{CB}X + XY + X^{CB}Y
daughter   daughter   son   son
       (carrier)   (color blind)

Thus, half the daughters of the carrier will also be carriers and half the sons will be color blind.

4. An F′ factor may be used in a similar way as transducing phages to map the bacterial genes carried by the F′ factor. It may also be usful in complementation tests of the genes carried on the F′ factor (for example, cis–trans tests, which cannot be made on haploid bacteria).

5. By matching overlapping regions, the following circular map is constructed.

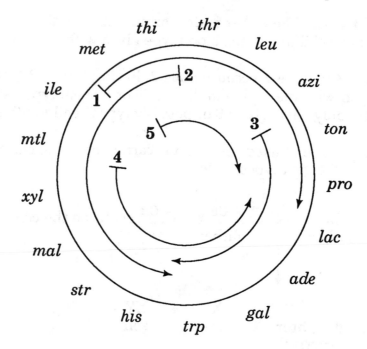

Chapter 28

NUCLEIC ACID STRUCTURES AND MANIPULATION

1. (a) **Cytosine** because hypoxanthine is a guanine that lacks its amino group.

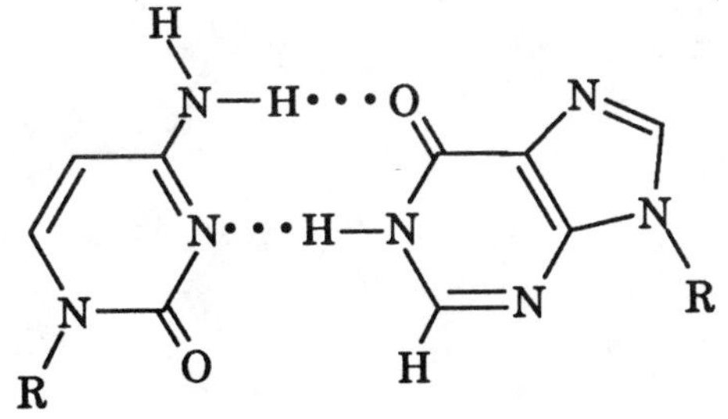

C Hypoxanthine

(b)

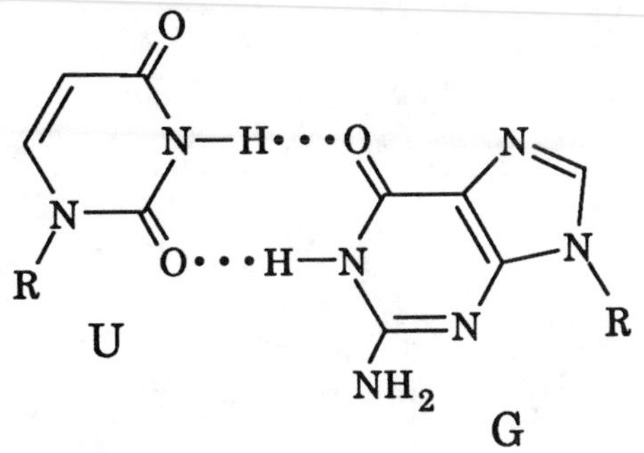

U G

(c)

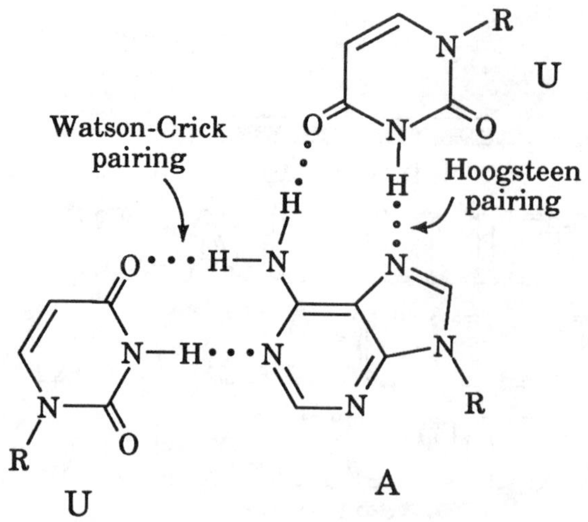

Watson-Crick pairing / Hoogsteen pairing

U / U / A

or

Watson-Crick pairing / Reversed Hoogsteen pairing

U / U / A

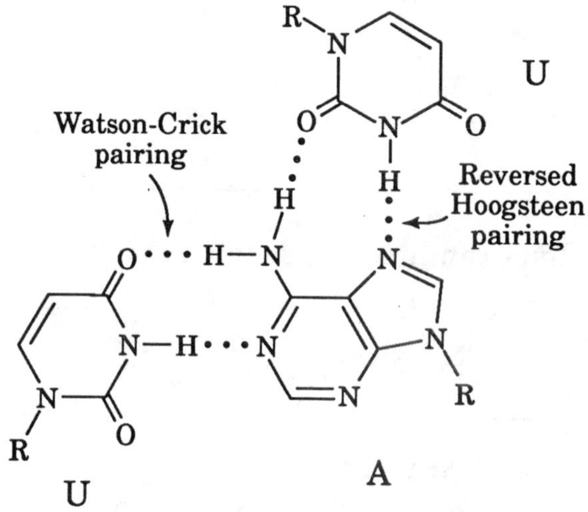

The U on the left might also form a reversed Hoogsteen pairing in either case.

(d)

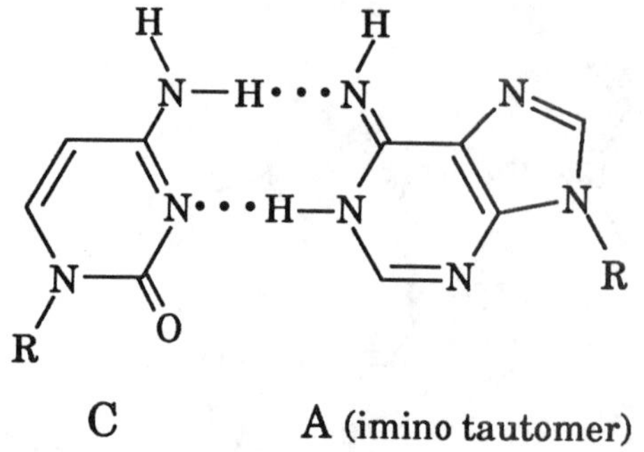

C   A (imino tautomer)

This has the same geometry as C•G or T•A. Hence, the replicated strand, in this case, has a C base pairing with A rather than T. In the next round of replication, this mutant C directs the incorporation of a G. Hence, the mutation is A → G on the parent strand (T → C in the complementary strand).

2. Since the average molecular mass of an amino acid residue is 110 D, the protein has $40,000/110 \approx 360$ residues. This is specified by 3 x 360 = 1080 nucleotides. Each base pair has an average molecular mass of 660 D. Hence, the molecular mass of the DNA is 660 x 1080 ≈ **700,000 D**. The length of this molecule, considering that B-DNA has a rise along its helix axis of 3.4 Å per base pair, is 3.4 Å x 1080 = **3700 Å = 0.37 μm.**

3. The only constraint on the relative ratios of the various dinucleotides in duplex DNA is that the amount of any particular dinucleotide must be equal to that of its complementary dinucleotide. Hence, **in a duplex that has antiparallel strands:**

   ApA = TpT, CpA = TpG, GpA = TpC, *etc.*

whereas **in a duplex with parallel strands:**

   ApA = TpT, CpA = GpT, GpA = CpT, TpA = ApT, *etc.*

4. (a)   The T_m would decrease because the phosphate charges would be less shielded from each other.

   (b)   The resulting shearing forces would break DNA into relatively small fragments. This would decrease the cooperativity of their melting and therefore broaden the melting curve.

   (c)   The adenine would compete with the adenine residues on the DNA for base pairing with the thymine — but not very effectively. Likewise, the adenine would compete rather poorly for base stacking interactions with the bases of the DNA. Both effects would slightly lower T_m.

   (d)   Heating and rapid cooling would separate the complementary DNA strands. Consequently, any residual hyperchromism would arise from the melting of small inter- and intramolecularly base paired segments. These would have a very broad and shallow melting curve which, of course, would be superimposable on that of native structure once T_m had been surpassed.

   (e)   This nonpolar solvent would disrupt the hydrophobic forces stabilizing DNA and hence lower its T_m.

5. At high pH, the ring N—H protons of uracil and guanine ionize. This disrupts the base pairing which, together with the mutual repulsions from the additional negative charges, results in strand separation.

6. At high [Na$^+$], there is little water available to promote hydrophobic bonding because most water molecules are involved in solvating the ions in the solution. Consequently, the structure of double-helical DNA, which is largely stabilized by hydrophobic forces, is destabilized by increasing [Na$^+$] as is indicated by its decreasing T_m.

7. The bonds opposite atoms C(2′) and C(3′) in the ribose ring both contain O(1′). Since O(1′) has no extra-annular substituents, none of the substituents to the ribose ring will be eclipsed if either C(2′) or C(3′) are out of the plane of the other ring atoms. This minimizes the steric interference between ring substituents.

8. **Type I is closed circular duplex DNA, type II is a nicked circular duplex DNA, and type III is linear duplex DNA.** Type I DNA has such a broad melting curve and a high T_m because its strands cannot separate from each other without breaking a covalent bond (its ΔS of melting is unusually small). The bases therefore tend to remain paired above the T_m for the equivalent linear DNA.

   The slightly greater sedimentation coefficient of type I in comparison with type II DNA indicates that type I is supercoiled which makes it more compact than type II. The relaxed circles of type II DNA are more compact than type III, the linear duplex, and hence type II DNA sediments faster than type III.

   At pH 13, which denatures duplex DNA, type III DNA undergoes strand separation to yield linear single strands of 16S. Type II DNA also undergoes strand separation to yield a 16S linear single strand and a slightly more compact 18S single-stranded circle. Type I DNA denatures but its strands cannot unwind from each other (L remains constant). However, the stiff double-helical structure has collapsed so that denatured type I DNA is a highly compact entity. This accounts for its 53S sedimentation coefficient.

9. In the B-DNA → Z-DNA conversion, a right-handed helix with one turn per 10.5 base pairs converts to a left-handed helix with one turn per 12 base pairs. Since a right handed duplex helix has a positive twist, the twist decreases. Thus,

$$\Delta T = -\left(\frac{100}{10.5}\right) + \left(\frac{-100}{12}\right) = -17.9 \text{ turns}$$

   The linking number must remain constant ($\Delta L = 0$) since no covalent bonds are broken. Hence, the change in writhing number is:

$$\Delta W = -\Delta T = 17.9 \text{ turns}$$

10. **The enzyme has no effect on the supercoiling of DNA** since cleaving the C(2′)—C(3′) bond of ribose does not sever the sugar-phosphate chain of DNA.

11. In a duplex DNA of 40% G+C there is a 1/5 chance each that any given base would be G or C and a 3/10 chance each that it will A or T.

*Taq*I recognizes TCGA so that its recognition sequence is to be randomly expected every

$$\frac{1}{\frac{3}{10}\times\frac{1}{5}\times\frac{1}{5}\times\frac{3}{10}}=278 \text{ base pairs}$$

Hence SV40 should have about 5243/278 ≈ **19 *Taq*I sites.**

*Eco*RII recognizes CC$\overset{A}{\underset{T}{}}$GG. Its recognition sequence should randomly occur every

$$\frac{1}{\frac{1}{5}\times\frac{1}{5}\times\frac{6}{10}\times\frac{1}{5}\times\frac{1}{5}}=1042 \text{ base pairs}$$

SV40 is therefore expected to contain 5243/1042 ≈ **5 *Eco*RII sites.**

*Pst*I recognizes CTGCAG. Its recognition sequence randomly occurs every

$$\frac{1}{\frac{1}{5}\times\frac{3}{10}\times\frac{1}{5}\times\frac{1}{5}\times\frac{3}{10}\times\frac{1}{5}}=6944 \text{ base pairs}$$

SV40 should have about 5243/6944 ≈ **1 *Pst*I site.**

*Hae*II recognizes PuGCGCPy. There is 1/2 chance that any site in duplex DNA is a purine (Pu) and a 1/2 chance that it is pyrimidine (Py). Hence, the *Hae*II recognition sequence should randomly occur every

$$\frac{1}{\frac{1}{2}\times\frac{1}{5}\times\frac{1}{5}\times\frac{1}{5}\times\frac{1}{5}\times\frac{1}{2}}=2500 \text{ base pairs}$$

SV40 should have about 5243/2500 ≈ **2 *Hae*II sites.**

12. Naked nicked circular duplex DNA cannot maintain its supercoils because the intact sugar phosphate chain can swivel about the bonds opposite the nick. The observation that nicking does not abolish supercoiling in the single DNA molecule of the bacterial chromosome indicates that its proteins somehow prevent this swiveling or at least prevent its effects from being transmitted the entire length of the DNA molecule. For example, if the DNA in the chromosome consists of a series of loops rigidly held by protein as diagrammed below, nicking one loop would not affect the supercoiling in neighboring loops.

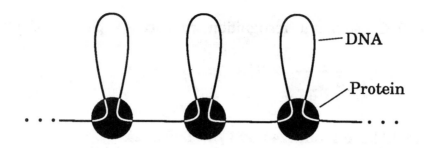

13. Blunt ends: *Alu*I, *Fnu*DI, and *Hae*III.

    Isoschizomers: (*Fnu*DI, *Hae*III).

    Isocaudamers: (*Bam*HI, *Bgl*II), (*Hpa*III, *Taq*I), and (*Sal*I, *Xho*I).

14. The sizes of the fragments in each digest total 5.4 kb so this must be the size of the plasmid. Neither *Eco*RI nor *Sal*I alone affect the size of the plasmid but together they fragment it. Hence, the plasmid must be circular with *Eco*RI and *Sal*I each making one restriction cut which the *Eco*RI + *Sal*I data indicate are 2.2 kb apart. By examining the remaining data in the Table, the following restriction map can be made where the *Eco*RI site is arbitrarily set at 0 kb and the *Sal*I site is placed 2.2 kb away from the *Eco*RI site in the clockwise direction.

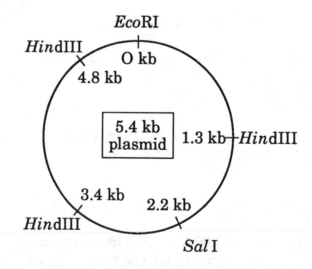

15. Since the 3′ end is ^{32}P-labeled in this case, the fragments lengthen towards the 5′ end. Hence, the sequence, as read from the bottom of the gel, is in the 3′ → 5′ direction. In the customary 5′ → 3′ direction, it is:

```
           10            20            30            40            50
            |             |             |             |             |
5′  ACAGCTATTGCTTTGAGATTCTGGAGCGGCGGTAATTTTGTATAGAATTT ··· 3′
```

16. A 5000 bp DNA fragment represents a fraction $f = 5000/(13{,}500 \times 1000) = 3.70 \times 10^{-4}$ of the yeast genome. According to equation [28.7],

$$N = \ln(1-P)/\ln(1-f)$$

For $P = 0.90$:

$$N = \ln(1 - 0.90)/(1 - 3.70 \times 10^{-4}) = \mathbf{6216}$$

For $P = 0.99$:

$$N = \ln(1 - 0.99)/(1 - 3.70 \times 10^{-4}) = \mathbf{12{,}432}$$

For $P = 0.999$:

$$N = \ln(1 - 0.999)/(1 - 3.70 \times 10^{-4}) = \mathbf{18{,}648}$$

17. The scheme would not work because the λ DNA without its central section is too small to be packaged into phage heads. Consequently, such DNA would not be proliferated.

---

1. (a) Inducible Z, no Y.

   (b) Constitutive Z, inducible Y.

   (c) Constitutive Z and Y.

   (d) No Z or Y production.

2. (a) Uninducible, no enzyme produced.

   (b) Constitutive production of Z (even an I^S repressor cannot bind to O^c).

   (c) Uninducible, no enzyme produced (the gene product of I^S binds to both O-operators).

3. The residual β-galactosidase in uninduced wild-type *E. coli* catalyzes the formation of the physiological inducer allolactose. Hence, the presence of lactose does not induce Z⁻ *E. coli* to synthesize lac enzymes. Residual galactoside permease permits lactose to enter an uninduced wild-type *E. coli* cell. Consequently, lactose in the medium of a Y⁻ cell fails to enter the cell to induce *lac* enzyme synthesis.

4. IPTG is not degraded by *E. coli* as is allolactose so that the amount of inducer present is easily controlled.

5. The top strand is the anti-sense strand:

$$\begin{array}{ccc} -30 & -10 & -1 \\ | & | & | \end{array}$$

5' CAA...CAC⎡TTTACA⎤GCG...TGA⎡TATGAT⎤GCGCCCC⎡G⎤CTT...ATA 3'

　　　　 −35 Region　　　　　Pribnow box　　Start point

6. Since rifamycin-sensitive holoenzyme is unable to initiate transcription, it becomes permanently bound to a promoter. Rifamycin-resistant holoenzyme is unable to initiate transcription at such blocked promoters.

7. Let x be the probability that a base is incorrectly transcribed. Then the probability that the base is correctly transcribed is $1 - x$ and the probability that all 4026 bases encoding the β subunit of RNA polymerase are correctly transcribed is $P = (1 - x)^{4026}$.

Thus,

$$\text{for } x = 0.0001, \qquad P = (1 - 0.0001)^{4026} = \mathbf{0.668}$$

$$\text{for } x = 0.001, \qquad P = (1 - 0.001)^{4026} = \mathbf{0.018}$$

$$\text{for } x = 0.01, \qquad P = (1 - 0.01)^{4026} = \mathbf{2.59 \times 10^{-18}}$$

8. There are 14 symmetry related base pairs in each half of the lac operator (see Figure 29-17). There is a probability of 1/4 that any two base pairs are randomly related by symmetry. Hence, the total probability of finding all 14 pairs of base pairs agreeing by random chance is $(1/4)^{14} = \mathbf{3.7 \times 10^{-9}}$.

9. DNA gyrases (Section 28-5C) cause the negative supercoiling of DNA which tends to unwind its double helix. This most probably aids in the formation of the open initiation complex that RNA polymerase makes with the promoter. In catabolite sensitive operons, the binding of CAP-cAMP, it is thought, further aids in this unwinding process. Inhibition of DNA gyrase therefore makes CAP-cAMP's job much harder.

10. In the absence of ribosomes, the 1•2 stem may form. This pre-empts the formation of the 2•3 stem so that the 3•4 terminator stem may form. Hence, the *trp* operon is not transcribed in the absence of active ribosomes.

11. In eukaryotes, transcription takes place in the nucleus and translation occurs in the cytoplasm. Hence, in eukaryotes, ribosomes are never in contact with nascent mRNAs which is an essential aspect of the attenuation mechanism in prokaryotes.

12. The 15-mer should base pair with segment 1 of the attenuator. This permits the 2•3 stem to form which, in turn, prevents the formation of the 3•4 terminator stem. Hence, the

presence of the 15-mer should increase the rate of attenuation in an *in vitro* system. However, a mutation in segment 2 that destabilizes the 2•3 stem would permit the 3•4 stem to form regardless of the state of association of segment 1. Consequently, such a mutation would render the 15-mer unable to affect the attenuation process.

13. *relA⁻* mutants have defective stringent factor so that they cannot make ppGpp. This substance signals the deficiency of amino acids and stimulates the transcription of biosynthetic operons such as the *his* and *trp* operons. The absence of ppGpp therefore greatly reduces the transcription of such operons.

14. In wild-type *E. coli*, RNA processing begins before a transcript has been completely synthesized. Hence, primary rRNA transcripts never actually exist in these organisms.

Chapter 30
TRANSLATION

1. Nitrous acid converts amino groups to carbonyl groups. Hence, the reaction is

Guanine $\xrightarrow{\text{HNO}_2}$ **Xanthine**

The product, xanthine, base pairs with C in normal Watson-Crick geometry (although it has 2 rather than 3 hydrogen bonds),

Xanthine   **Cytosine**

so that reaction with HNO_2 is not mutagenic for guanine.

2. The DNA antisense strand specifies the complementary RNA which, in turn, specifies the polypeptide:

mRNA: 5'- A GCU AUG UGC CAG AGA GCU CAA UAG UCA GA

Polypeptide: initiate Cys Gln Arg Ala Gln stop

3. The wild-type and mutant polypeptides differ only in the middle of their corresponding sequences. It therefore appears that the mutations were caused by two nearly opposing frameshift mutations.

By comparing the codons specifying the various corresponding amino acids, the following sequences can be deduced.

$$
\text{Wild type}
\begin{cases}
\text{Cys} & \text{Glu} & \text{Asp} & \text{His} & \text{Val} & \text{Pro} & \text{Gln} & \text{Tyr} & \text{Arg} \\
\text{UG}_C^U & \text{GA}_G^A & \underline{\text{G}}\text{AC} & \text{CAU} & \text{GUC} & \text{CCA} & \text{CAG} & \text{UA}_C^U & \text{CGX} \\
 & & & & & & & & \text{AG}_G^A
\end{cases}
$$

$$
\text{Mutant}
\begin{cases}
 & & & & & & & & \text{AG}_G^A \\
\text{UG}_C^U & \text{GA}_G^A & \text{ACC} & \text{AUG} & \text{UCC} & \text{CAC} & \text{AG}_{\underline{C}}^U & \text{UA}_C^U & \text{CGX} \\
\text{Cys} & \text{Glu} & \text{Thr} & \text{Met} & \text{Ser} & \text{His} & \text{Ser} & \text{Tyr} & \text{Arg}
\end{cases}
$$

One mutation is a deletion of the G(underlined) in the codon specifying the wild-type Asp. The other mutation is the insertion of a U or C preceeding the leading U of the wild-type Tyr or the insertion of a U following the leading U of this Tyr. The mutated peptide must have little specific function since changing its entire sequence does not drastically affect the function of the protein.

4. Transitions change one purine to another or one pyrimidine to another. Hence, two consecutive transitions at the same position results in no net change.

Transversions change purines to pyrimidines and vice versa. Two consecutive transversions at the same position results in no net change.

Frame shift mutations are deletions or insertions. A deletion of one or more nucleotides followed by their replacement (or vice versa) results in no net change.

One of these types of mutations cannot undo the changes caused by another type.

5. An *amber* mutation can be achieved by any of the point mutations XAG, UXG or UAX→ UAG (*amber*). The XAG codons specify Gln, Lys and Glu; the UXG codons specify Leu, Ser and Trp; and the UAX codons that are not stop codons both specify Tyr. Hence, some of the codons specifying these amino acids can undergo a point mutation to the amber codon.

6. The two C-terminal peptides have a common Ser-Lys-Tyr-Arg sequence but that of Hb Constant Spring continues past this normal Hbα C-terminus. Evidently, the native Hb stop codon:

$$... \text{ UCC AAA UAC CGU UAA GCU GGA } ...$$

$$\text{Ser} \quad \text{Lys} \quad \text{Tyr} \quad \text{Arg} \quad \text{stop}$$

must be altered to that specifying Gln. Here the UAA stop codon has been altered to specify Gln (CAA or CAG). Apparently, the U of the stop codon has undergone a transition to C so as to specify Gln. Then, the succeeding GCU will specify Ala and GGA will specify Gly as occurs in the mutant.

7. From Table 30-2: The 2 amino acids specified by only one codon each require a tRNA. The 12 amino acids that are specified by PQX where X may be either purine or else either pyrimidine can each be specified by a single tRNA according to the wobble pairing rules (Table 30-4). Ile, which is specified by AUY, where Y = U, C or A, also requires a single tRNA. However, the 8 amino acids that are specified by RSZ, where Z = U, C, A or G, must each have 2 tRNAs specifying them (Leu, Arg and Ser are specified by both PQX and RSZ). Finally, an initiator tRNA is required. Thus,

$$2 + 1 + 12 + 2 \times 8 + 1 = 32 \text{ tRNAs are minimally required.}$$

8. The wobble pairings (Table 30-4) not in Figure 30-21a are C•I (which closely resembles C•G) and U•I (which resembles U•G).

9. HNO$_2$ deaminates amino groups thereby converting A → I, C → U, and G → xanthine (nonmutagenic). The Gly codon GGX (Table 30-2) is recognized by the Gly anticodon XCC. This, in turn, is specified by the tRNA^{Gly} DNA sequence GGX or its complement XCC. A change of X results in another X so that such a mutant tRNA^{Gly} still recognizes only Gly codons. G is not mutagenized by HNO$_2$. However, the DNA sequence XCC may be converted to XUC, XCU, or XUU whose complements specify anticodons of the same sequence. The *amber* codon UAG is not recognized by any of these anticodons. Hence, **your colleague is mistaken.**

10.

*su*1: UAG is recognized by **CUA**.

The Ser codons are UCX and AG^U_C which are recognized by the anticodons XGA and ^A_GCU. Thus, to match CUA, the original tRNA^{Ser} anticodon must have been CGA and the mutation was **2nd G → U**.

*su*2: UAG is recognized by **CUA**.

The Gln codons are CA^A_G which are recognized by ^U_CUG. Thus, the tRNA^{Gln} that mutated must have had the anticodon sequence CUG and the mutation was **3rd G → A**.

*su*3: UAG is recognized by **CUA**.

The Tyr codons are UA^U_C which are recognized by ^G_AUA. Thus, the mutation was **1st ^G_A → C.**

*su*4: According to wobble rules, UA^A_G is recognized by **UUA**.

The Tyr codons are UA^U_C which are recognized by ^G_AUA. Thus, the mutation was **1st ^G_A → U.**

*su*5: According to wobble rules, UA^A_G is recognized by **UUA**.

The Lys codons are AA^A_G which are recognized by ^U_CUU. Thus, the tRNA^{Lys} that mutated must have had the anticodon sequence UUU and the mutation was **3rd U → A.**

*su*6: UAA is recognized by **UUA**.

The Leu codons UU^A_G and CUX are recognized by ^U_CAA and XAG. Thus, the tRNA^{Leu} that mutated must have had the anticodon sequence UAA and the mutation was **2nd A → U.**

*su*7: UAA is recognized by **UUA**.

The Gln codons are CA^A_G which are recognized by ^U_CUG.
Thus, the tRNA^{Gln} that mutated must have had the anticodon sequence UUG and the mutation was **3rd G → A.**

UGA-2: UGA is recognized by **UCA**.

The Trp codon is UGG which is recognized by CCA. Thus, the mutation was **1st C → U.**

11. The ribosome contains 52 proteins + 3 RNAs.

   A minimum of 31 noninitiating tRNAs and their 20 cognate aminoacyl-tRNA synthetases are required.

   $tRNA_f^{Met}$ is also required (it is charged by the aminoacyl-tRNA synthetase for Met).

   Initiation requires 3 factors: IF-1, IF-2, and IF-3.

   Elongation requires 3 factors: EF-Tu, EF-Ts, and EF-G.

   Termination requires 3 factors: RF-1, RF-2, and RF-3.

   mRNA is also required.

   Thus, the total number of macromolecules is $52 + 3 + 31 + 20 + 1 + 3 + 3 + 3 + 1$
   = **117 different macromolecules.**

12. Oligonucleotides containing Shine-Dalgarno sequences bind to the 3' end of 16S RNA thereby competing with the Shine-Dalgarno sequences of the translational initiation regions of prokaryotic mRNA. Eukaryotic mRNAs and 18S RNAs contain no such sequences and are therefore not affected by these oligonucleotides.

13. m^7GTP is a component of the 5'-cap structure of eukaryotic mRNAs (Figure 29-31). Recognition of the cap plays an important role in eukaryotic ribosomal initiation. Thus, m^7GTP interferes with this process. However, prokaryotic mRNAs are uncapped and prokaryotic ribosomes do not recognize caps. Thus, prokaryotic ribosomal initiation is unaffected by m^7GTP.

14. Addition of unlabeled leucine would permit the synthesis of each labeled hemoglobin polypeptide to continue through its termination and release. Hence, each of these polypeptides would be radioactively labeled over a small region but the position of this region would vary randomly among these polypeptides since their synthesis is not synchronized.

15.

		Start	Lys	Pro	Ala

$$5'\text{- AGGAGCUX}_{\sim4} \quad {}^A_G\text{UG} \quad \text{AA}{}^A_G \quad \text{CCX} \quad \text{GCX-}$$

Shine-Dalgarno sequence.
3-10 base pairs with G•U's
allowed

Gly	Thr	Glu	Asn	Ser	stop
GGX	ACX	GA${}^A_G$	AA${}^U_C$	UCX	UAA
				or	UAG - 3'
				AG${}^U_C$	UGA

16.

Shine-Dalgarno
sequence

	Start	Cys	Gln	Ser	Arg	Met

5'- CUGAUAAGGGAUUUAAAUU AUG UGU CAA UCA CGA AUG-

Leu	Ile	Glu	Ala	Pro	stop	stop

CUA AUC GAG GCU CCA UAA UAA CACUUCGAC-3'

17. Assuming the leading fMet residue is cleaved from the mature polypeptide as it often is:

Initiation requires 1 ATP (a GTP is energetically equivalent to an ATP).

Elongation requires 2ATP for each of the 100 elongation cycles.

Termination requires 1 ATP.

The synthesis of each aminoacyl-tRNA requires 1 ATP. A total of 101 aminoacyl-tRNA's are required: the leading Met-tRNA$_f$Met + 100 others.

Thus, the number of ATPs = 1 + 100 x 2 + 1 + 101 = **303ATP** (if the leading fMet remained on the mature polypeptide, this number would be 300 since one less aminoacyl-tRNA and one less elongation step would be required to form the 100-residue polypeptide).

18. A plausible editing mechanism for aminoacyl-tRNA synthetase is that the amino acid binding site sterically screens out side chains larger than that of its cognate amino acid. The proofreading site then binds and hydrolyzes the resulting aminoacyl-tRNA only if its amino acid side chain is smaller than that of the cognate amino acid. Gly, being the smallest amino acid, cannot have such a proofreading step but also does not need it since

all other amino acids are sterically excluded, a highly discriminating process, in the initial binding step.

19. Fixmycin apparently inhibits translocation. Since dipeptides are formed, it must not inhibit initiation, aminoacyl-tRNA binding in the A site, or transpeptidation. There is also no indication that it inhibits termination (although there is likewise no information to the contrary).

20. When heme is present in significant amounts there must be insufficient globin present to sequester it. The presence of heme both inhibits protein degradation and stimulates protein synthesis. This is a feedback mechanism that matches the amount of heme to the amount of globin in the reticulocyte.

21. The gene for the enzyme should have a leading sequence coding for a signal peptide of 15 to 30 mostly hydrophobic residues. This causes the nascent polypeptide to be secreted through the cell membrane.

Chapter 31
DNA REPLICATION, REPAIR, AND RECOMBINATION

1. Pol I has two structurally independent domains: the larger domain contains its polymerase and $3' \rightarrow 5'$ exonuclease functions and the smaller domain contains its $5' \rightarrow 3'$ exonuclease function. A point mutation or insertion/deletion that alters the catalytic site on the larger domain need not affect the smaller domain. Alternatively, since the smaller domain is N-terminal, a frameshift or nonsense mutation in the larger subunit could abolish its activity without affecting the smaller subunit.

2. The $5' \rightarrow 3'$ exonuclease function of Pol I is essential for the replication of DNA because it removes the RNA primers and replaces them with DNA. Hence, a mutation that totally inactivates this function at low temperatures is lethal.

3. Type I topoisomerases relieve negative supercoils in DNA (Section 28-5C). However, the unwinding of DNA, as occurs at the replication fork, generates positive supercoils in a circular duplex. Thus, if replication is to proceed, negative supercoils must be generated, not relaxed, so as to absorb the positive supercoils. This is done by DNA gyrase (topoisomerase II). Hence, topoisomerase I does not participate in DNA replication.

4. DNA polymerization results in the formation of base pairs. Since a back reaction should be the exact reverse of a forward reaction, pyrophosphorolysis, the back reaction of DNA polymerization, should act only on base paired 3'-terminal nucleotides. Consequently, there must be two forms of the enzyme-DNA complex. That which is base paired favors synthesis or pyrophosphorolysis, whereas that which is unpaired favors hydrolysis.

Hence, the enzyme must have two at least partially separate active sites for these activities.

5. (a) *dnaB* encodes a subunit of the primosome which is required for the initiation of Okazaki fragments. Inactivation of DnaB protein stops lagging strand synthesis when the temperature is raised and thus stops the progression of the replication fork. This would be lethal.

(b) *dnaE* encodes the α subunit of pol III, the DNA replicase. Its inactivation would prevent DNA replication. This would be lethal.

(c) *dnaG* encodes primase. Its inactivation stops replication for the reasons stated in Part (a). This would be lethal.

(d) *lig* encodes DNA ligase. Its inactivation would result in the synthesis of short segments of DNA rather than continuous strands. This would be lethal.

(e) *polA* encodes Pol I. Its inactivation would prevent the excision of the RNA primers and excision repair. Hence, the DNA would be fragmented with RNA leaders on the resulting Okazaki fragments. This would be lethal.

(f) *rep* encodes Rep protein which helps unwind the double helix ahead of the replication fork. However, helicase II also participates in this process. Hence, inactivating Rep would slow but not prevent DNA replication (Section 31-2C).

(g) *ssb* encodes single-strand binding protein (SSB) which prevents separated single strands ahead of the replication fork from reannealing. The inactivation of SSB would stop replication. This would be lethal.

(h) *recA* encodes RecA which mediates general recombination and the SOS response. Inactivation of RecA would render bacteria unable to undergo normal recombination and hypersensitive to massive DNA damage.

6. Okazaki fragments are 1000 to 2000 nucleotides in length. The *E. coli* chromosome consists of 4,000,000 bp (Table 28-2). Consequently, during chromosomal replication, *E. coli* must generate 2000 to 4000 Okazaki fragments.

7. Replication of DNA, which requires 40 minutes, must commence 60 minutes before the corresponding cell division. For a 25 minute cell generation time, the following diagram can be drawn in analogy with Figure 31-22.

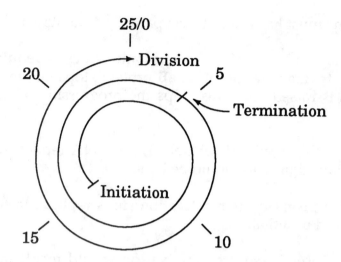

Since a new round of DNA replication must commence every 25 minutes, the succeeding round of replication is also occurring between 25 and 40 minutes after a given round has commenced. After termination (at 40 minutes), each chromosome will have only one fork until the 3rd replication round is initiated at 50 minutes after first round initiation. However, the number of replication forks double at every level and replication is bidirectional in *E. coli*. Hence, as the following drawing indicates, the number of replication forks varies between 2 and 6.

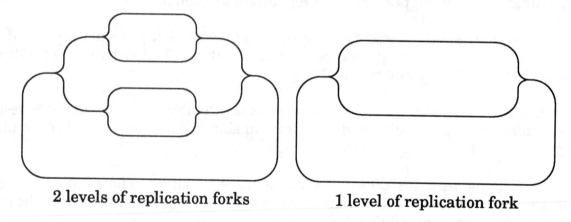

2 levels of replication forks   1 level of replication fork

For 80 minute doubling, initiation will occur once per generation at 60 minutes before cell division:

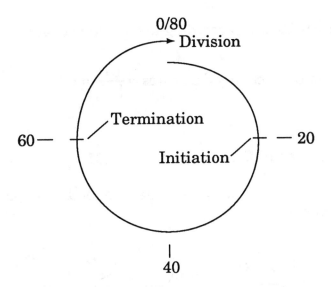

The replication fork only exists between 20 and 60 minutes from the beginning of a generation. Since replication is bidirectional, there are either 0 or 2 replication forks in the chromosome.

8. DNA polymerase can only extend a chain with a free 3′-OH group. Consequently, in lagging strand synthesis, the primer closest to the 3′ end of the template could never be replaced by DNA. Thus, in all of the mechanisms considered in Chapter 31, the 3′ ends of the parental strands could not be replicated (viruses with linear duplex DNAs, as well as eukaryotes, whose chromosomes are also linear, have a variety of specialized processes for replicating the ends of linear duplexes).

9. The human genome consists of 2,900,000,000 bp (Table 28-2) which is also its number of purine nucleotides. There are 10,000 spontaneous depurinations per day in the human genome so that the rate constant for first order decay is:

$$k = \frac{10,000 \ \text{purines/genome-day}}{2,900,000,000 \ \text{purines/genome}} = 3.45 \times 10^{-6} \ \text{day}^{-1}$$

According to equation [13.5], the half-time for a first order process is

$$t_{1/2} = \frac{0.693}{k}$$

so that

$$t_{1/2} = \frac{0.693}{3.45 \times 10^{-6} \ \text{day}^{-1}} = 2.01 \times 10^{5} \ \text{days} = \frac{2.01 \times 10^{5} \ \text{days}}{365 \ \text{days/year}} = 550 \ \text{years}$$

In 25 years = 25 x 365 = 9125 days, according to equation [13.4b],

$$\text{Fraction of unchanged purines} = \frac{[A]}{[A]_o} = e^{-kt} = e^{-3.45 \times 10^{-6} \times 9125} = 0.969$$

so that

$$\text{Fraction depurinated} = 1 - 0.969 = \mathbf{0.031}$$

10. O^6-methylguanine can form a doubly hydrogen bonded base pair of normal Watson-Crick geometry with thymine:

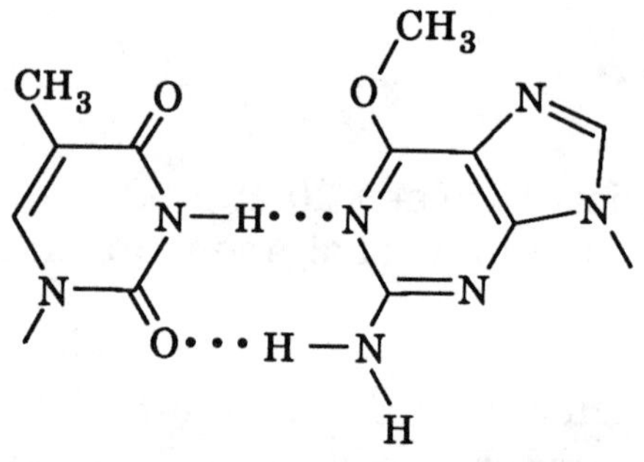

T O^6 -methyl G

Hence, DNA replication in which such a base pair forms results in a C•G → T•A mutation.

11. When 5-methylcytosine residues deaminate, they form thymine residues:

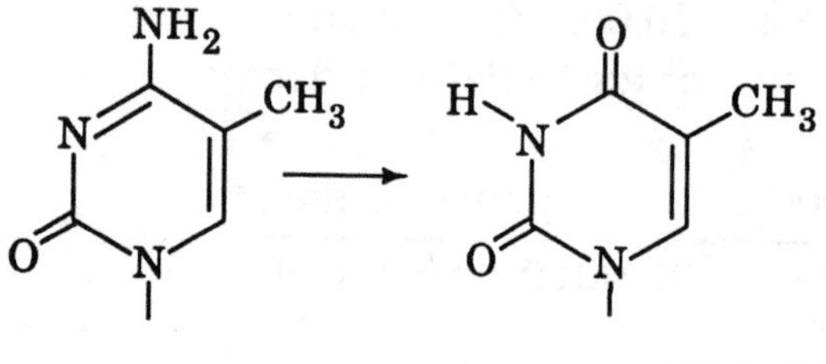

5-Methyl-C   T

Since thymine is a normal DNA base, the repair systems are unable to determine whether such a T or its opposing G is the mutated base. Hence, the deamination of 5-methyl-C, a relatively frequent event, is only correctly repaired half the time it occurs.

12. The methylation pattern in a cell is "heritable" in that it is copied by maintanence methylases from parent to progeny DNA strands. Exposure, even briefly, to 5-azacytosine alters this pattern, both because the nonmethylatable 5-azacytosine may be substituted for C in DNA and because 5-azacytosine is a DNA-methylase inhibitor. Since the pattern of methylation apparently controls gene expression in many organisms, a change of the methylation pattern will alter the phenotype of these organisms.

13. Chi structures arise from two identical plasmids in the process of crossing over ("figure-8" structures) that have been treated with a restriction enzyme. The plasmids therefore have the same restriction sites so that the arms extending from the crossover point of a chi structure are identical in pairs.

14. Transposons are flanked by inverted repeats which, on a single strand, are complementary. Consequently, in a single-stranded circle containing a transposon sequence, the inverted repeats will be internally base paired so as to form a stem-and-double-loop structure.

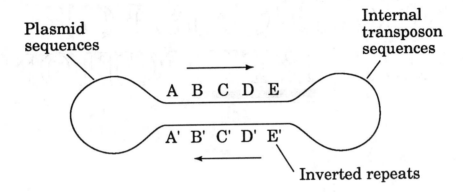

15. A composite transposon consists of a central region flanked by two IS-like units which, in turn, are each flanked by inverted repeats:

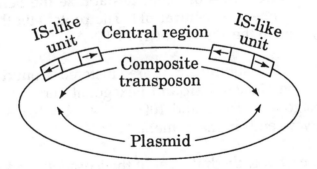

Thus, the plasmid has the same relationship to the IS-like units as does the transposon's central region; that is, it is flanked by these IS-like units with their flanking inverted repeats. The entire plasmid, with the flanking IS-units, may therefore be transposed rather than the composite transposon. Indeed, the plasmid with the flanking IS-like units is a composite transposon.

Chapter 32

VIRUSES: PARADIGMS FOR CELLULAR FUNCTIONS

1. At acidic pH's, the anomalously basic carboxyl groups will be protonated, thereby eliminating the electrostatic repulsions that destabilize the helical conformation with respect to that of the protohelix at neutral pH. The protohelix therefore converts to the helical conformation, even in the absence of TMV RNA.

2. An icosadeltahedron has $60T$ subunits that are clustered around 12 pentagonal vertices with the remainder being clustered around hexagonal vertices. Thus, it has $(60T - 5 \times 12)/6 = 10(T - 1)$ hexagonal vertices and $10(T - 1) + 12 = 10T + 2$ vertices of any kind. This latter number always ends in the numeral "2".

3. The triangular face of each icosahedral facet of the icosadeltahedron is subdivided into 9 triangles as in:

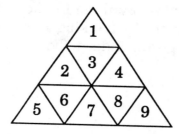

4. It is impossible to grow phages with a lethal mutation and thus one could not obtain the defective protein in order to study its effect on phage assembly.

5. B-DNA can be considered to be a cylinder of around 20 Å in diameter that has a length of 3.4 Å per base pair. The λ DNA, which consists of 48,502 bp, occupies a volume of about

$$\pi \left(\frac{20}{2}\text{Å}\right)^2 \times 48{,}502 \text{ bp} \times 3.4 \text{ Å/bp} = 5.2 \times 10^7 \text{ Å}^3 = 5.2 \times 10^4 \text{ nm}^3$$

The icosahedral phage head, which is 55 nm in diameter, can be considered to be roughly spherical. Hence, its volume is around

$$\frac{4}{3}\pi \left(\frac{55 \text{ nm}}{2}\right)^3 = 8.7 \times 10^4 \text{ nm}^3$$

Thus, considering the fact that the volume of its coat protein must occupy a significant fraction of the phage head's volume, **the volume of the DNA and the volume contained by the capsid are roughly equal.**

6. A plaque is a phage colony that has descended from a single progenitor. A clear plaque indicates that all the bacteria in it have been lysed by the virus. In a turbid plaque, only some of the bacteria have been lysed. In a phage λ infection, the phages initially follow a lytic life cycle. However, when the phages become so numerous that each host cell is multiply infected or when the bacteria have nearly exhausted the available nutrients, the phages assume a lysogenic life cycle and thus coexist with their hosts. Since lysogeny protects a bacterium from superinfection by additional phages, not all the bacteria in the plaque are lysed in a phage λ infection. Consequently, their plaques are cloudy.

7. Consulting Figure 32-32 (red-lettered half of each duplex), the aggregate base composition at each position of the half subsites is:

base/site	1	2	3	4	5	6	7	8	9
No. A's	1	12	3		9				1
No. T's	9		6		2	1	4	2	1
No. G's							1	9	5
No. C's	2		3	12	1	11	7	1	5
Consensus	T	A	T	C	A	C	C	G	$\frac{G}{C}$

$$\left.\begin{array}{l}\text{Consensus sequence}\\\text{for half-site}\end{array}\right\} \quad \begin{array}{l} \text{T A T C A C C G } \frac{G}{C} \\ \text{A T A G T G G C } \bullet \end{array}$$

8. The interactions between repressor dimers bound at o_{R1} and o_{R2} must physically preclude the repressor bound at o_{R2} from associating with that bound at o_{R3}. For

example, the repressor bound at o_{R2} may have to lean to the "right" to contact repressor at o_{R1} so that repressor at o_{R2} cannot simultaneously contact repressor at o_{R1} and o_{R3}. If, however, o_{R1} cannot bind repressor, repressor bound at o_{R2} will be free to associate with repressor at o_{R3}.

9. The α3 helices of repressor and Cro proteins both bind in the major grooves of their initial target operator DNAs. Thus, the α3 helix provides the major contacts of these proteins with their operators. The hybrid protein should therefore bind to DNA as does 434 Cro protein and consequently exhibit Cro's protection pattern.

10. The probability that the second RNA chosen differs from the first RNA chosen is 7/8; the probability that the third chosen differs from the first two is 6/8, *etc.* Hence, the probability that the first 8 RNA's chosen will all be different is:

$$P = \frac{7}{8} \times \frac{6}{8} \times \frac{5}{8} \times \frac{4}{8} \times \frac{3}{8} \times \frac{2}{8} \times \frac{1}{8} = \frac{7!}{8^7} = 0.0024$$

11. Viroids are highly self-complementary and therefore, over long stretches, are essentially double-stranded. It is this double-stranded character that probably permits viroids to act as templates for RNA polymerase II.

Chapter 33
EUKARYOTIC GENE EXPRESSION

1. The packing ratio of DNA is defined as the ratio of its contour length to the length of its container. This ratio is maximal when the container is a sphere (if the container were a cube, for example, its longest dimension, its body diagonal, would be larger than the diameter of a sphere of equal volume).

   The volume, V, occupied by a 10^6 bp DNA is a 20 Å in diameter cylinder that is 3.4×10^6 Å long.

   $$V = \pi \left(\frac{d}{2}\right)^2 l = 3.14 \times \left(\frac{20}{2}\right)^2 \times 3.4 \times 10^6 \text{ Å}^3 = 1.07 \times 10^9 \text{ Å}^3$$

   The radius of sphere of this volume ($V = 4\pi r^3/3$) is

   $$r = \left[\frac{3V}{4\pi}\right]^{1/3} = \left[\frac{3 \times 1.07 \times 10^9}{4 \times 3.14}\right]^{1/3} = 635 \text{ Å}$$

   $$\text{Packing ratio} = \frac{\text{contour length}}{\text{diameter of sphere}} = \frac{3.4 \times 10^6 \text{ Å}}{2 \times 635 \text{ Å}} = 2680$$

For a 10^9 bp DNA, repeating the calculations:

$$r = \left[\frac{3 \times 1.07 \times 10^{12}}{4 \times 3.14}\right]^{1/3} = 6350 \text{ Å}$$

$$\text{Packing ratio} = \frac{3.4 \times 10^9 \text{ Å}}{2 \times 6350 \text{ Å}} = 268,000$$

2. If the DNA double helical winding was unaffected by supercoiling around the nucleosome, the SV40 minichromosome would be supercoiled by nearly –2 superhelical turns per nucleosome (left-handed toroidal turns, by definition, have negative supercoils; Section 28-5A). The observed –1 superhelical turns per nucleosome implies that the missing superhelical turns are taken up by the duplex helix. This, in turn, implies that the duplex DNA in a nucleosome is overwound in its relaxed state by about one turn per nucleosome (which is $360°/146$ bp $\approx 2.5°/$bp).

3. Histones have a great tendency to form nonspecific aggregates. The presence of the polyanion polyglutamate apparently electrostatically shields the histones from each other and thereby allows them to specifically aggregate with DNA to form nucleosomes. Thus, polyglutamate, much like the acidic protein nucleoplasmin, acts as a "molecular chaperone" in the formation of nucleosomes.

4. The DNA molecule consists of 60% repeated sequences of complexity $x = 400$ and 40% unique sequences with $x = 400,000$. Hence, the $C_0t_{1/2}$ values of these two types of sequences should differ by a factor of 1000. There should be little, if any, difference between the C_0t curves of DNA sheared to lengths of 1000 bp and 100 bp because the rate of collision between complimentary sequences is the same in both cases. Hence, in both cases, the C_0t curve has the shape:

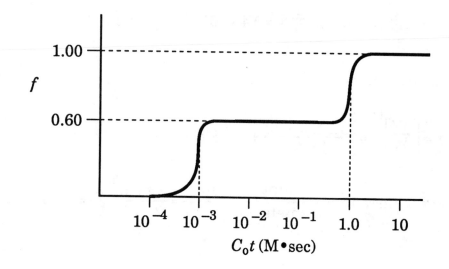

5. Single-strand nucleases would remove the non-base paired loop of the foldback structures:

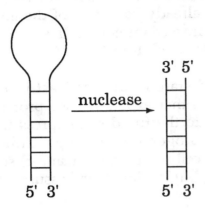

Since the sequences of these foldback structures vary in their complexity as much as the rest of the DNA sample from which they are taken (at least to the limit of their length), cleaved and denatured foldback structures can take varied amounts of time to renature.

6. (a) The rDNA transcripts (rRNA) are used directly to form ribosomes. However, many copies of a ribosomal protein can be translated from a single copy of ribosomal mRNA. Thus, in a sense, the ribosomal proteins are "amplified" in the normal course of their synthesis.

   (b) Assuming that the rate of rRNA synthesis is proportional to the number of rDNA copies (that is, that the availability of template DNA is the limiting factor in its transcription), in the absence of gene amplification, rRNA synthesis should occupy 1500 x 2 months = **250 years.**

7. Hb Kenya most probably arose in a manner similar to Hb Lepore (Figure 33-37): by an unequal crossing-over between the $^A\gamma$-gene and the corresponding position of the β-gene.

8. The band is indicative of a region in the DNA made DNase I hypersensitive by the binding of Sp1.

9. Calico cats are genetic mosaics in which some patches of skin grow yellow fur as specified by one X chromosome and the remaining patches grow black fur as specified by the second X chromosome (Section 33-3A). Normal XY male cats, having only one X chromosome, cannot be calico cats. However, genetically abnormal XXY male cats can be calico cats since one of their two X chromsomes in each cell is inactivated to form a Barr body, just as in normal females.

10. The *esc* gene is apparently a maternal-effect gene. Thus, the proper distribution of the *esc* gene product in the fertilized egg, which is maternally specified, is sufficient to permit normal embryonic development regardless of the embryo's genotype.

11. Cell transformation results from several genetic changes in a cell. Thus, a single oncogene supplied to an otherwise normal cell will be insufficient to transform it. However, an immortalized cell already has some of the genetic changes necessary for transformation (malignant cells are also immortal). In such cells, the additional oncogene may be all they require to complete their transformation.

12. If the cancer cell's transformed state results, at least in part, from the absence of a functional anti-oncogene such as that expressing Rb protein, and if the chromosomes that the normal cell contribute to the fused cell express that anti-oncogene, then the fused cell will have a nontumorogenic phenotype. This is because anti-oncogene products suppress uncontrolled cell proliferation (cancer) so that cells requiring such a gene product for normal growth, but lacking it, will assume the cancerous state.

Chapter 34
MOLECULAR PHYSIOLOGY

1. The only symptom of vitamin K deficiency in humans is poor blood clotting caused by the lack of Gla residues in various clotting factors. Vitamin K is a cofactor for the conversion of Glu to Gla in these proteins.

2. Citrate, which has three carboxylate groups, chelates and thus sequesters Ca^{2+}. Since Ca^{2+} is an essential cofactor in most steps of the blood clotting cascade (Figure 34-1), blood clotting is thereby inhibited.

3. Both antihemophilic factor (VIII) and Christmas factor (IX) are required for clotting. They are absent in the blood of individuals with hemophilias A and B. A mixture of these two bloods contains both factors (although at reduced concentrations) so that it clots.

4. (a) Precipitation requires a minimally divalent antibody to cross-link antigens with multiple antigenic sites. An Fab fragment contains only one antigen-binding site and therefore cannot cross-link antibodies.

   (b) A hapten is a small molecule that specifically binds to antibodies raised against hapten-protein complexes. Since haptens have only one antigenic site, antibodies cannot cross-link them so that precipitates cannot form.

   (c) When either antigen or antibody is in excess, the resultant antibody-antigen complex will not be cross-linked. Rather, the agent in excess will monovalently bind to all the available binding sites on the agent not in excess. Consequently, no large insoluble aggregates can form.

5. The antigenic determinents on a native protein are formed by patches on its surface and are therefore a consequence of the protein's three-dimensional structure. Denaturing the protein usually disrupts these antigenic determinants (epitopes).

6. (a) An antibody's antigen-combining site is geometrically complementary to a portion on the surface of the antigen. An antibody raised against the first antibody may have an antigen-combining site that is complementary to that of the first antibody and thus resembles the first antibody's antigen, that is, the enzyme surface. If the original antigenic site is the enzyme's active site, the second antibody's antigen-combining site may geometrically resemble the enzyme's active site and consequently bind the enzyme's substrate.

    (b) Enzymes tend to bind the transition state of the reaction they catalyze with greater affinity than the substrate itself (transition state stabilization; Section 14-1F). An antibody that binds a transition state analog of a reaction would likewise preferentially bind the reaction's transition state and thereby catalyze the reaction.

7. *T*-Cells recognize their cognate foreign antigen only when that antigen is in complex with a self-MHC protein. The MHC proteins of another individual apparently often resemble such a complex, which is not unexpected since they are allelic proteins.

8. The binding of antibody to a foreign antigen, whether it's on a living or dead foreign invader, triggers various processes such as inflammation and opsonization (phagocytosis of foreign particles) which cause many of the symptoms of infection.

9. Since the tense muscle maintains a constant length, its myosin heads cannot "walk" up the actin filaments in a concerted fashion as they do during muscle contraction. Rather, they "dither", that is, a given myosin head "walks" up a thin filament for a short distance thereby generating tension and then relaxes thereby releasing the thin filament. The myosin head "walking" consumes ATP as it does during a normal contraction but the energy is dissipated as heat since the muscle as a whole is prevented from doing thermodynamic work.

10. In the absence of ATP, each myosin head in its low energy configuration is bound to a neighboring actin filament. It will not release the actin filament until ATP binds to it and consequently thick and thin filaments form a rigid cross-linked array — the rigor complex.

11. Tropomyosin inhibits the binding of myosin to muscle thin filaments and presumably does the same with the closely similar microfilaments. Microfilaments with bound tropomyosin therefore cannot interact with myosin thick filaments to contract.

12. Colchicine inhibits the formation of microtubules but not their dissolution. Nevertheless,

since *in vivo* microtubules are simultaneously assembling at one end and disassembling at the other, the inhibition of the assembly process, for example, by colchicine, allows the disassembly process to predominate so that previously existing microtubules eventually disappear.

13. (a) Removal of the thyroid gland prevents the secretion of thyroid hormones, T_3 and T_4, which function to stimulate metabolism. The absence of the thyroid gland consequently depresses the metabolism rate, thereby permitting thyroidectomized rats to survive longer on their stored nutrients than normal rats.

    (b) A pituitary tumor may cause excessive synthesis of ACTH which stimulates the release of adrenocortical hormones.

    (c) Injury to the pituitary may prevent the release of vasopressin which inhibits the kidney from excreting water. Lack of vasopressin thereby greatly increases the rate of urination and the consequent loss of water leads to unquenchable thirst.

    (d) Sex-organ derived tissues are often highly responsive to sex steroids which are secreted by the gonads and the adrenal cortex. Removal of these glands therefore removes a powerful growth stimulant for these tissues.

14. Guanylylimido diphosphate binds to the $G_{s\alpha}$ subunit of G protein as does GTP but cannot be hydrolyzed by this protein to GDP. The binding of this GTP inhibitor to $G_{s\alpha}$ therefore all but irreversibly activates $G_{s\alpha}$, thereby inappropriately activating adenylate cyclase. In the presence of a $G_{i\alpha}$ system, the adenylate cyclase will similarly be abnormally inactivated.

15. (a) In the presence of tetrodotoxin, the Na^+ channels are blocked so that $P_{Na} = 0$. Consequently, the Goldman equation ([34.9]) reduces to:

$$\Delta\Psi = \frac{RT}{F}\ln\left\{\frac{P_K[K^+(\text{out})] + P_{Cl}[Cl^-(\text{in})]}{P_K[K^+(\text{in})] + P_{Cl}[Cl^-(\text{out})]}\right\}$$

so that using the data in Table 34-6,

$$\Delta\Psi = \frac{8.314 \text{ J}\bullet\text{K}^{-1}\text{mol}^{-1} \times (273+25)\text{ K}}{96,494 \text{ C}\bullet\text{mol}^{-1}}\ln\left\{\frac{5\times10^{-7}\times4 + 1\times10^{-8}\times116}{5\times10^{-7}\times139 + 1\times10^{-8}\times4}\right\}$$

$$\Delta\Psi = -0.079 \text{ J}\bullet\text{C}^{-1} = -0.079 \text{ V} = -79 \text{ mV (hyperpolarized)}$$

(b) Cs^+ blocks the K^+ channel from the inside and hence $P_K = 0$ in its presence. Therefore,

$$\Delta\Psi = \frac{RT}{F}\ln\left\{\frac{P_{Na}[Na^+(out)] + P_{Cl}[Cl^-(in)]}{P_{Na}[Na^+(in)] + P_{Cl}[Cl^-(out)]}\right\}$$

$$= \frac{8.314\ J\bullet K^{-1}mol^{-1} \times (273+25)\ K}{96,494\ C\bullet mol^{-1}}\ln\left\{\frac{5\times10^{-9}\times145 + 1\times10^{-8}\times116}{5\times10^{-9}\times12 + 1\times10^{-8}\times4}\right\}$$

$$\Delta\Psi = 0.075\ J\bullet C = 0.075\ V = +75mV \qquad \text{(depolarized)}$$

Tetrodotoxin's blocking of the Na^+ channel prevents membrane depolarization and hence prevents the triggering of action potentials. Blocking the K^+ channel depolarizes the membrane so that an action potential cannot develop.

16. Nerve impulses cannot propagate in the reverse direction because a membrane patch immediately upstream from a membrane patch undergoing an action potential has just undergone an action potential itself and therefore is in its refractory period. The upstream membrane patch will not undergo another action potential until the action potential under consideration has moved so far downstream that is is unable to trigger an action potential on the membrane patch in question.

17. Decamethonium is a nonhydrolyzable acetylcholine analog. It therefore locks the ACh receptor in the open position making skeletal muscles unresponsive to nerve impulses and thereby relaxing them.